# ANATOMY & PHYSIOLOGY COLORING WORKBOOK

## A Complete Study Guide

### SEVENTH EDITION

**Elaine N. Marieb, R.N., Ph.D.**

Holyoke Community College

Benjamin
Cummings

San Francisco    Boston    New York
Cape Town    Hong Kong    London    Madrid    Mexico City
Montreal    Munich    Paris    Singapore    Sydney    Tokyo    Toronto

Publisher: Daryl Fox

Managing Editor, Editorial: Kay Ueno

Associate Project Editor: Mary Ann Murray

Managing Editor, Production: Wendy Earl

Production Editor: David Novak

Composition: Greene Design

Text Designer: Carolyn Deacy

Cover Designer: f. tani hasegawa

**Library of Congress Cataloging-in-Publication Data**

Marieb, Elaine Nicpon
    Anatomy and physiology coloring workbook : a complete study guide /
Elaine N. Marieb.-- 7th ed.
        p. cm.
    ISBN 0-8053-5903-6
        1. Human anatomy--Problems, exercises, etc. 2. Human physiology--
Problems, exercises, etc. I. Title.

QM32 .M34 2002
611'.0076--dc21

2002024659

Benjamin
Cummings

ISBN 0-8053-5903-6

4 5 6 7 8 9 10—MAL—06 05 04 03

www.aw.com/bc

# THE BENJAMIN CUMMINGS SERIES IN HUMAN ANATOMY AND PHYSIOLOGY

## Two-Semester Human Anatomy and Physiology

Elaine N. Marieb
*Human Anatomy and Physiology,*
Fifth Edition

Elaine N. Marieb
*Anatomy and Physiology,*
First Edition

## One-Semester Human Anatomy and Physiology

Elaine N. Marieb
*Essentials of Human Anatomy and Physiology,* Seventh Edition

## Supplements for Human Anatomy and Physiology

Glenn Bastian
*Anatomy Flash Cards*

Patricia J. Donnelly and George A. Wistreich
*Laboratory Manual for Anatomy and Physiology with Cat Dissections,*
Fifth Edition

Mark Lafferty and Samuel Pinella
*A.D.A.M.® Interactive Anatomy Student Lab Guide,* Second Edition

Elaine N. Marieb
*Human Anatomy and Physiology Laboratory Manual, Main Version,*
Sixth Edition Update

Elaine N. Marieb
*Human Anatomy and Physiology Laboratory Manual, Cat Version,*
Seventh Edition Update

Elaine N. Marieb
*Human Anatomy and Physiology Laboratory Manual, Fetal Pig Version,*
Seventh Edition Update

Elaine N. Marieb
*Essentials of Human Anatomy and Physiology Laboratory Manual,*
Second Edition

Dennis Strete
*A Color Atlas of Histology*

Rose Leigh Vines
*Human Anatomy Video Series*
*Student Workbook for A.D.A.M.®*

## Electronic Supplements

A.D.A.M.®—Benjamin Cummings
*InterActive Physiology® 7-System Suite*
    Muscular System
    Nervous System I
    Nervous System II
    Cardiovascular System
    Respiratory System
    Urinary System
    Fluid, Electrolyte, and
        Acid/Base Balance
*Anatomy Practice*

Glenn Bastian
*An Illustrated Review of Anatomy and Physiology: Systems Series*

Peter Zao and Timothy Stabler
*PhysioEx v4.0, CD-ROM Version*
*PhysioEx v4.0, Web Version*

## Human Anatomy

Elaine N. Marieb and Jon Mallatt
*Human Anatomy,* Third Edition Update

## Supplements for Human Anatomy

Glenn Bastian
*Clip Art of the Human Body*

Robert A. Chase
*The Bassett Atlas of Human Anatomy*

Victor Eroschenko
*Laboratory Manual for Human Anatomy with Cadavers,*
Second Edition

Wynn Kapit and Larry Elson
*The Anatomy Coloring Book,*
Third Edition

Steve Langjahr and Bob Brister
*Coloring Atlas of Human Body,*
Second Edition

Elaine N. Marieb
*Human Anatomy Laboratory Manual with Cat Dissections,*
Third Edition

Rose Leigh Vines
*Human Anatomy Video Series*

## Human Physiology

William Germann and Cindy L. Stanfield
*Principles of Human Physiology*

Richard L. Malvin, Michael D. Johnson, and Gary Malvin
*Concepts of Human Physiology*

## Supplements for Human Physiology

A.D.A.M.®—Benjamin Cummings
*InterActive Physiology® 7-System Suite*
   Muscular System
   Nervous System I
   Nervous System II
   Cardiovascular System
   Respiratory System
   Urinary System
   Fluid, Electrolyte, and
      Acid/Base Balance

Wynn Kapit, Robert I. Macey, and Esmail Meisami
*The Physiology Coloring Book,*
Second Edition

To order any of our products, call customer service at **(800) 824-7799**, send a fax to **(800) 835-5327**, call **(201) 767-4900** outside the U.S., or visit your campus bookstore. For further information, visit us at **www.awl.com/bc** on the World Wide Web.

# PREFACE

Although never a simple task, the study of the human body is always fascinating. The Seventh Edition of the *Anatomy & Physiology Coloring Workbook* serves as a review and reinforcement tool to help health professional and life-science students master the basic concepts of human anatomy and physiology.

## Scope

Although this book reviews the human body from microscopic to macroscopic levels (that is, topics range from simple chemistry and cells to body organ systems), it is not intended to be encyclopedic. In fact, to facilitate learning, this workbook covers only the most important and useful aspects of human anatomy and physiology. Pathophysiology is briefly introduced with each system so that students can apply their learning. Where relevant, clinical aspects (for example, muscles used for injection sites, the role of ciliated cells in protection of the respiratory tract, and reasons for skin ulcer formation) are discussed. To encourage a view of the human body as a dynamic and continually changing organism, developmental aspects of youth, adulthood, and old age are included.

## Learning Aids

As in previous editions, multiple pedagogical devices are used throughout the book to test comprehension of key concepts. The integration of a traditional study guide approach with visualization and coloring exercises is unique. The variety of exercises demands learning on several levels, avoids rote memorization, and helps maintain a high level of interest.

The exercises include completion from a selection of key choices, matching terms or descriptions, and labeling diagrams. Elimination questions require the student to discover the similarities or dissimilarities among a number of structures or objects and to select the one that is not appropriate. Correctable true/false questions add a new dimension to the more traditional form of this exercise. Also, students are asked to provide important definitions. In the completion sections, the answer lines are long enough so that the student can write in either the key letter or the appropriate term. Both responses are provided in the answer section.

Coloring exercises are a proven motivating, effective approach to learning. Each illustration has been carefully prepared to show sufficient detail for learning without students becoming bored with coloring. There are over 100 coloring exercises distributed throughout the text that should prove valuable to all students. Students who are visually oriented will find these exercises particularly beneficial. When completed, the color diagrams provide an ideal reference and review tool.

Visualization exercises are a truly unique feature of this book. With the exception of the introductory chapter on terminology, each chapter contains an Incredible Journey. Students are asked to imagine themselves in miniature, traveling within the body through various organs and systems. These visualization exercises are optional, but they often summarize chapter content, allowing students to assimilate what they have learned in unusual and amusing ways.

As the final exercises of each chapter, thought-provoking "At the Clinic" questions challenge students to apply their newly acquired knowledge to clinical situations.

## Acknowledgments

To those educators, colleagues, and students who have provided feedback and suggestions during the preparation of all seven editions of this workbook, I am sincerely grateful. The staff at Benjamin Cummings has continuously supported my efforts to turn out a study tool that will be well received and beneficial to both educator and student audiences. I wish to thank the entire team at Wendy Earl Productions and particularly David Novak for his diligent work as production editor on this edition.

# INSTRUCTIONS FOR THE STUDENT—
# HOW TO USE THIS BOOK

Dear Student,

The *Anatomy & Physiology Coloring Workbook* has been created particularly for you. It is the outcome of years of personal attempts to find and create exercises helpful to my own students when they study and review for a lecture test or laboratory quiz.

Although I never cease to be amazed at how remarkable the human body is, I would never try to convince you that studying it is easy. The study of human anatomy and physiology has its own special terminology. It requires that you become familiar with the basic concepts of chemistry to understand physiology, and often (sadly) it requires rote memorization of facts. It is my hope that this workbook will help simplify your task. To make the most of the exercises, read these instructions carefully before starting work.

*Labeling and Coloring.* Some of these questions ask you only to label a diagram, but most also ask that you do some coloring of the figure. You can usually choose whichever colors you prefer. Soft colored pencils are recommended so that the underlying diagram shows through. Most figures have several parts to color, so you will need a variety of colors—18 should be sufficient. In the coloring exercises, you are asked to choose a particular color for each structure to be colored. That color is then used to fill in both a color-coding circle found next to the name of the structure or organ, and the structure or organ on the figure. This allows you to identify the colored structure quickly and by name in cases where the diagram is not labeled. In a few cases you are given specific coloring instructions to follow.

*Matching.* Here you are asked to match a key term denoting a structure or physiologic process with a descriptive phrase or sentence. Because you must write the chosen term in the appropriate answer blank, the learning is more enduring.

*Completion.* You select the correct term to answer a specific question, or you fill in blanks to complete a sentence. In many exercises, some terms are used more than once and others are not used at all.

*Definitions.* You are asked to provide a brief definition of a particular structure or process.

*True or False.* One word or phrase is underlined in a sentence. You decide if the sentence is true as it is written. If not, you correct the underlined word or phrase.

***Elimination.*** Here you are asked to find the term that does not "belong" in a particular grouping of related terms. In this type of exercise, you must analyze how the various terms are similar to or different from the others.

***Visualization.*** "The Incredible Journey" is a special type of completion exercise, found in every chapter except the first one. For this exercise, you are asked to imagine that you have been miniaturized and injected into the body of a human being (your host). Anatomical landmarks and physiological events are described from your miniaturized viewpoint, and you are then asked to identify your observations. Although this exercise is optional, my students have found them fun to complete and I hope you will too.

***At the Clinic.*** "At the Clinic" sections end each chapter and ask you to apply your newly acquired knowledge to clinical situations.

Each exercise has complete instructions, which you should read carefully before beginning the exercise. When there are multiple instructions, complete them in the order given.

At times it may appear that information is duplicated in the different types of exercises. Although there is some overlap, the understandings being tested are different in the different exercises. Remember, when you understand a concept from several different perspectives, you have mastered that concept.

I sincerely hope that the *Anatomy & Physiology Coloring Workbook* challenges you to increase your knowledge, comprehension, retention, and appreciation of the structure and function of the human body.

Good luck!

*Elaine Marieb*

Elaine Marieb
Benjamin Cummings Publishing
1301 Sansome Street
San Francisco, CA 94111

# CONTENTS

## *Chapter 1  The Human Body: An Orientation*  1

An Overview of Anatomy and Physiology  1

Levels of Structural Organization  2

Maintaining Life  7

Homeostasis  8

The Language of Anatomy  8

At the Clinic  14

## *Chapter 2  Basic Chemistry*  15

Concepts of Matter and Energy  15

Composition of Matter  16

Molecules, Chemical Bonds, and Chemical Reactions  18

Biochemistry: The Composition of Living Matter  21

Incredible Journey: A Visualization Exercise for Biochemistry  27

At the Clinic  28

## *Chapter 3  Cells and Tissues*  29

Cells  29

    Overview  29

    Anatomy of a Generalized Cell  30

    Cell Physiology  32

Body Tissues  38

    Tissue Repair  43

Developmental Aspects of Cells and Tissues  44

Incredible Journey: A Visualization Exercise for the Cell  45

At the Clinic  46

## *Chapter 4 Skin and Body Membranes* 49

Classification of Body Membranes 49

Integumentary System (Skin) 51

    Basic Structure and Function 51

    Appendages 53

    Homeostatic Imbalances of the Skin 55

Developmental Aspects of the Skin and Body Membranes 56

Incredible Journey: A Visualization Exercise for the Skin 56

At the Clinic 58

## *Chapter 5 The Skeletal System* 61

Bones—An Overview 61

Axial Skeleton 64

    Skull 64

    Vertebral Column 69

    Bony Thorax 72

Appendicular Skeleton 73

Bone Fractures 82

Joints 83

    Homeostatic Imbalances of Bones and Joints 85

Developmental Aspects of the Skeleton 85

Incredible Journey: A Visualization Exercise for the Skeletal System 86

At the Clinic 87

## *Chapter 6 The Muscular System* 89

Overview of Muscle Tissues 89

Microscopic Anatomy of Skeletal Muscle 91

Skeletal Muscle Activity 93

Muscle Movements, Types, and Names 96

Gross Anatomy of the Skeletal Muscles 98

    Muscles of the Head 98

    Muscles of the Trunk 100

    Muscles of the Hip, Thigh, and Leg 104

    Muscles of the Arm and Forearm 106

    General Body Muscle Review 107

Developmental Aspects of the Muscular System  112

Incredible Journey: A Visualization Exercise for the Muscular System  112

At the Clinic  113

## Chapter 7  The Nervous System  115

Organization of the Nervous System  116

Nervous Tissue—Structure and Function  116

Central Nervous System  122

    Brain  122

    Protection of the CNS—Meninges and Cerebrospinal Fluid  126

    Brain Dysfunctions  127

    Spinal Cord  128

Peripheral Nervous System  130

    Structure of a Nerve  130

    Cranial Nerves  132

    Spinal Nerves and Nerve Plexuses  133

    Autonomic Nervous System  134

Developmental Aspects of the Nervous System  136

Incredible Journey: A Visualization Exercise for the Nervous System  136

At the Clinic  138

## Chapter 8  Special Senses  141

The Eye and Vision  141

The Ear: Hearing and Balance  146

Chemical Senses: Smell and Taste  150

Developmental Aspects of the Special Senses  153

Incredible Journey: A Visualization Exercise for the Special Senses  154

At the Clinic  155

## Chapter 9  The Endocrine System  157

The Endocrine System and Hormone Function—An Overview  157

The Major Endocrine Organs  159

Other Hormone-Producing Tissues and Organs  164

Developmental Aspects of the Endocrine System  164

Incredible Journey: A Visualization Exercise for the Endocrine System  165

At the Clinic  166

### *Chapter 10  Blood*  167

Composition and Functions of Blood  167

Hemostasis  172

Blood Groups and Transfusions  173

Developmental Aspects of Blood  173

Incredible Journey: A Visualization Exercise for the Blood  174

At the Clinic  175

### *Chapter 11  The Cardiovascular System*  177

Cardiovascular System: The Heart  177

Cardiovascular System: Blood Vessels  184

Developmental Aspects of the Cardiovascular System  196

Incredible Journey: A Visualization Exercise for
the Cardiovascular System  197

At the Clinic  199

### *Chapter 12  The Lymphatic System and Body Defenses*  201

The Lymphatic System  201

Lymphatic Vessels  201

Lymph Nodes and Other Lymphoid Organs  203

Body Defenses  206

Nonspecific Body Defenses  206

Specific Body Defenses: The Immune System  209

Disorders of Immunity  218

Developmental Aspects of the Lymphatic System and Body Defenses  218

Incredible Journey: A Visualization Exercise for the Immune System  219

At the Clinic  221

### *Chapter 13  The Respiratory System*  223

Functional Anatomy  223

Respiratory Physiology  230

Respiratory Disorders  234

Developmental Aspects of the Respiratory System  235

Incredible Journey: A Visualization Exercise for
the Respiratory System  236

At the Clinic  237

## *Chapter 14   The Digestive System and Body Metabolism*   239

Anatomy of the Digestive System   239

Functions of the Digestive System   250

Nutrition and Metabolism   254

Developmental Aspects of the Digestive System   258

Incredible Journey: A Visualization Exercise for the Digestive System   259

At the Clinic   261


## *Chapter 15   The Urinary System*   263

Kidneys   264

    Location and Structure   264

    Nephrons, Urine Formation, and Control of Blood Composition   266

Ureters, Urinary Bladder, and Urethra   271

Developmental Aspects of the Urinary System   273

Incredible Journey: A Visualization Exercise for the Urinary System   274

At the Clinic   275


## *Chapter 16   The Reproductive System*   277

Anatomy of the Male Reproductive System   277

Male Reproductive Functions   280

Anatomy of the Female Reproductive System   283

Female Reproductive Functions and Cycles   285

Mammary Glands   290

Survey of Pregnancy and Embryonic Development   290

Developmental Aspects of the Reproductive System   295

Incredible Journey: A Visualization Exercise for
    the Reproductive System   296

At the Clinic   297


**Answers**   299

**A.D.A.M.® Correlations**   347

# The Human Body: An Orientation

Most of us have a natural curiosity about our bodies, and a study of anatomy and physiology elaborates on this interest. Anatomists have developed a universally acceptable set of reference terms that allows body structures to be located and identified with a high degree of clarity. Initially, students might have difficulties with the language used to describe anatomy and physiology, but without such a special vocabulary, confusion is bound to occur.

The topics in this chapter enable students to test their mastery of terminology commonly used to describe the body and its various parts, and concepts concerning functions vital for life and homeostasis. Body organization from simple to complex levels and an introduction to the organ systems forming the body as a whole are also covered.

## AN OVERVIEW OF ANATOMY AND PHYSIOLOGY

**1.** Match the terms in Column B to the appropriate descriptions provided in Column A. Enter the correct letter or its corresponding term in the answer blanks.

| Column A | Column B |
|---|---|
| _____ 1. The branch of biological science that studies and describes how body parts work or function | A. Anatomy |
| _____ 2. The study of the shape and structure of body parts | B. Homeostasis |
| _____ 3. The tendency of the body's systems to maintain a relatively constant or balanced internal environment | C. Metabolism |
| _____ 4. The term that indicates *all* chemical reactions occurring in the body | D. Physiology |

**2.** Circle all the terms or phrases that correctly relate to the study of *physiology*. Use a highlighter to identify those terms or phrases that pertain to the study of *anatomy*.

A. Measuring an organ's size, shape, and weight

B. Can be studied in dead specimens

C. Often studied in living subjects

D. Chemistry principles

E. Measuring the acid content of the stomach

F. Principles of physics

G. Observing a heart in action

H. Dynamic

I. Dissection

J. Experimentation

K. Observation

L. Directional terms

M. Static

# LEVELS OF STRUCTURAL ORGANIZATION

**3.** The structures of the body are organized into successively larger and more complex structures. Fill in the answer blanks with the correct terms for these increasingly larger structures.

Chemicals ⟶ _____ ⟶ _____ ⟶

_____ ⟶ _____ ⟶ Organism

**4.** Circle the term that does not belong in each of the following groupings.

1. Atom      Cell       Tissue       Alive        Organ

2. Brain        Stomach        Heart      Liver         Epithelium

3. Epithelium      Heart      Muscle tissue      Nervous tissue      Connective tissue

4. Human        Digestive system        Horse        Pine tree        Amoeba

**5.** Using the key choices, identify the organ systems to which the following organs or functions belong. Insert the correct letter or term in the answer blanks.

*Key Choices*

A. Cardiovascular

B. Digestive

C. Endocrine

D. Integumentary

E. Lymphatic/Immune

F. Muscular

G. Nervous

H. Reproductive

I. Respiratory

J. Skeletal

K. Urinary

_____ 1. Rids the body of nitrogen-containing wastes

_____ 2. Is affected by the removal of the thyroid gland

_____ 3. Provides support and levers on which the muscular system can act

_____ 4. Includes the heart

_____ 5. Protects underlying organs from drying out and mechanical damage

_____ 6. Protects the body; destroys bacteria and tumor cells

_____ 7. Breaks down foodstuffs into small particles that can be absorbed

_____ 8. Removes carbon dioxide from the blood

_____ 9. Delivers oxygen and nutrients to the body tissues

_____ 10. Moves the limbs; allows facial expression

_____ 11. Conserves body water or eliminates excesses

_____ 12. Provides for conception and childbearing

_____ 13. Controls the body with chemicals called hormones

_____ 14. Is damaged when you cut your finger or get a severe sunburn

**6.** Using key choices from Exercise 5, choose the organ system to which each of the following sets of organs belongs. Enter the correct letter or term in the answer blanks.

_____ 1. Blood vessels, heart

_____ 2. Pancreas, pituitary, adrenal glands

_____ 3. Kidneys, bladder, ureters

_____ 4. Testis, vas deferens, urethra

_____ 5. Esophagus, large intestine, rectum

_____ 6. Breastbone, vertebral column, skull

_____ 7. Brain, nerves, sensory receptors

**7.** Figures 1–1 to 1–6, on pages 4–6, represent the various body organ systems. First identify and name each organ system by labeling the organ system under each illustration. Then select a different color for each organ and use it to color the coding circles and corresponding structures in the illustrations.

◯ Blood vessels

◯ Heart

◯ Nasal cavity

◯ Lungs

◯ Trachea

**Figure 1–1**                                    **Figure 1–2**

◯ Brain

◯ Spinal cord

◯ Nerves

◯ Kidneys

◯ Ureters

◯ Bladder

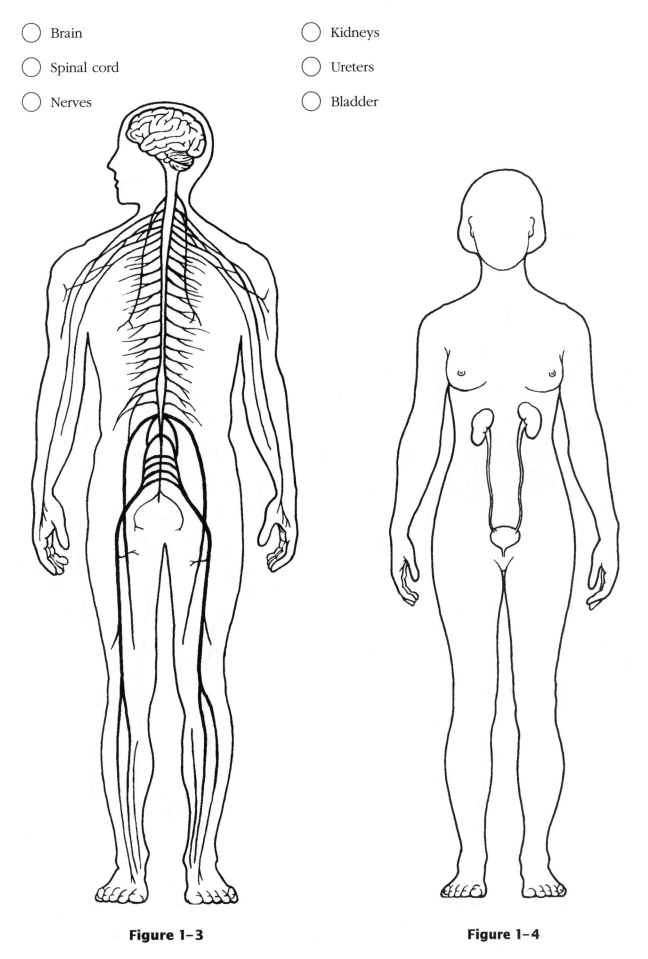

**Figure 1-3**

**Figure 1-4**

○ Stomach      ○ Esophagus      ○ Ovaries

○ Intestines   ○ Oral cavity    ○ Uterus

**Figure 1–5**                    **Figure 1–6**

# MAINTAINING LIFE

**8.** Match the terms pertaining to functional characteristics of organisms in Column B with the appropriate descriptions in Column A. Fill in the answer blanks with the appropriate letter or term.

**Column A**

**Column B**

_____ 1. Keeps the body's internal environment distinct from the external environment

A. Digestion

B. Excretion

_____ 2. Provides new cells for growth and repair

C. Growth

_____ 3. Occurs when constructive activities occur at a faster rate than destructive activities

D. Maintenance of boundaries

_____ 4. The tuna sandwich you have just eaten is broken down to its chemical building blocks

E. Metabolism

_____ 5. Elimination of carbon dioxide by the lungs and elimination of nitrogenous wastes by the kidneys

F. Movement

_____ 6. Ability to react to stimuli; a major role of the nervous system

G. Responsiveness

H. Reproduction

_____ 7. Walking, throwing a ball, riding a bicycle

_____ 8. All chemical reactions occurring in the body

_____ 9. At the cellular level, membranes; for the whole organism, the skin

**9.** Using the key choices, correctly identify the survival needs that correspond to the following descriptions. Insert the correct letter or term in the answer blanks.

*Key Choices*

A. Appropriate body temperature        C. Nutrients        E. Water

B. Atmospheric pressure        D. Oxygen

_____ 1. Includes carbohydrates, proteins, fats, and minerals

_____ 2. Essential for normal operation of the respiratory system and breathing

_____ 3. Single substance accounting for over 60% of body weight

_____ 4. Required for the release of energy from foodstuffs

_____ 5. Provides the basis for body fluids of all types

_____ 6. When too high or too low, physiological activities cease, primarily because molecules are destroyed or become nonfunctional

## HOMEOSTASIS

**10.** The following statements refer to homeostatic control systems. Complete each statement by inserting your answers in the answer blanks.

| | |
|---|---|
| _____ 1. | There are three essential components of all homeostatic control mechanisms: control center, receptor, and effector. The _(1)_ senses changes in the environment and responds by sending information (input) to the _(2)_ along the _(3)_ pathway. The _(4)_ analyzes the input, determines the appropriate response, and activates the _(5)_ by sending information along the _(6)_ pathway. When the response causes the initial stimulus to decline, the homeostatic mechanism is referred to as a _(7)_ feedback mechanism. When the response enhances the initial stimulus, the mechanism is called a _(8)_ feedback mechanism. _(9)_ feedback mechanisms are much more common in the body. |
| _____ 2. | |
| _____ 3. | |
| _____ 4. | |
| _____ 5. | |
| _____ 6. | |
| _____ 7. | |
| _____ 8. | |
| _____ 9. | |

## THE LANGUAGE OF ANATOMY

**11.** Complete the following statements by filling in the answer blanks with the correct term.

| | |
|---|---|
| _____ 1. | The abdominopelvic and thoracic cavities are subdivisions of the _(1)_ body cavity; the cranial and spinal cavities are parts of the _(2)_ body cavity. The _(3)_ body cavity is totally surrounded by bone and provides very good protection to the structures it contains. |
| _____ 2. | |
| _____ 3. | |

**12.** Circle the term or phrase that does not belong in each of the following groupings.

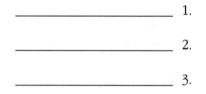

1. Transverse          Distal          Frontal          Sagittal

2. Lumbar          Thoracic          Antecubital          Abdominal

3. Sural          Brachial          Femoral          Popliteal

4. Epigastric          Hypogastric          Right iliac          Left upper quadrant

**13.** Select different colors for the *dorsal* and *ventral* body cavities. Color the coding circles below and the corresponding cavities in part A of Figure 1–7. Complete the figure by labeling those body cavity subdivisions that have a leader line. Complete part B by labeling each of the abdominal regions indicated by a leader line.

◯ Dorsal body cavity          ◯ Ventral body cavity

**A**

**B**

**Figure 1–7**

**14.** Select the key choices that identify the following body parts or areas. Enter the appropriate letter or corresponding term in the answer blanks.

### Key Choices

| | | | |
|---|---|---|---|
| A. Abdominal | E. Buccal | I. Inguinal | M. Pubic |
| B. Antecubital | F. Cervical | J. Lumbar | N. Scapular |
| C. Axillary | G. Femoral | K. Occipital | O. Sural |
| D. Brachial | H. Gluteal | L. Popliteal | P. Umbilical |

_____ 1. Armpit

_____ 2. Thigh region

_____ 3. Buttock area

_____ 4. Neck region

_____ 5. "Belly button" area

_____ 6. Genital area

_____ 7. Anterior aspect of elbow

_____ 8. Posterior aspect of head

_____ 9. Area where trunk meets thigh

_____10. Back area from ribs to hips

_____11. Pertaining to the cheek

**15.** Using the key terms from Exercise 14, correctly label all body areas indicated with leader lines on Figure 1–8.

In addition, identify the sections labeled A and B in the figure.

Section A:        _____

Section B:        _____

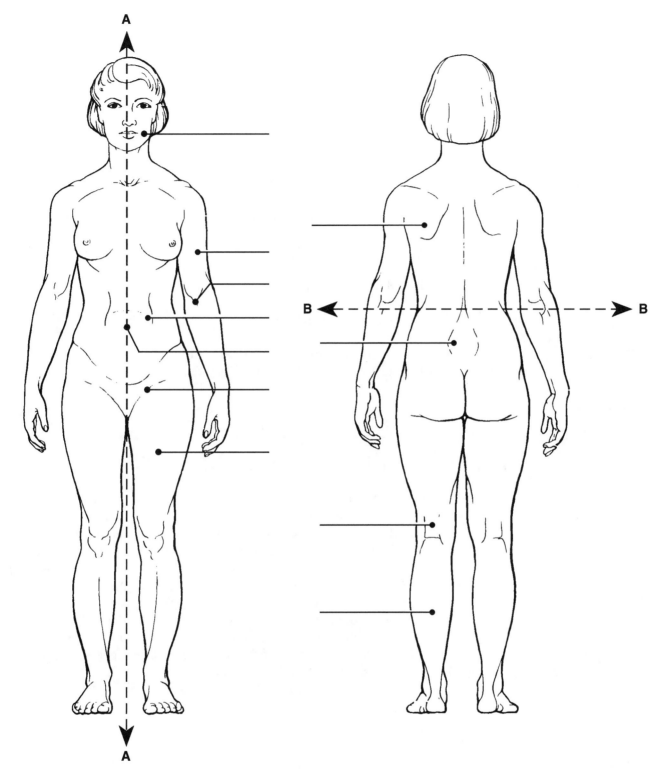

**Figure 1–8**

**16.** From the key choices, select the body cavities where the following surgical procedures would occur. Insert the correct letter or term in the answer blanks. Be precise: also select the name of the cavity subdivision if appropriate.

***Key Choices***

A. Abdominal    C. Dorsal    E. Spinal     G. Ventral

B. Cranial       D. Pelvic    F. Thoracic

_____ 1. Removal of the uterus, or womb

_____ 2. Coronary bypass surgery (heart surgery)

_____ 3. Removal of a serious brain tumor

_____ 4. Removal of a "hot" appendix

_____ 5. A stomach ulcer operation

**17.** Complete the following statements by choosing an anatomical term from the key choices. Enter the appropriate letter or term in the answer blanks.

***Key Choices***

A. Anterior    D. Inferior    G. Posterior    J. Superior

B. Distal      E. Lateral     H. Proximal    K. Transverse

C. Frontal     F. Medial      I. Sagittal

_____ 1.

_____ 2.

_____ 3.

_____ 4.

_____ 5.

_____ 6.

_____ 7.

_____ 8.

_____ 9.

_____ 10.

_____ 11.

In the anatomical position, the face and palms are on the __(1)__ body surface, the buttocks and shoulder blades are on the __(2)__ body surface, and the top of the head is the most __(3)__ part of the body. The ears are __(4)__ to the shoulders and __(5)__ to the nose. The heart is __(6)__ to the spine and __(7)__ to the lungs. The elbow is __(8)__ to the fingers but __(9)__ to the shoulder. In humans, the dorsal surface can also be called the __(10)__ surface; however, in four-legged animals, the dorsal surface is the __(11)__ surface.

_____ 12.
_____ 13.
_____ 14.
_____ 15.

If an incision cuts the heart into right and left parts, the section is a __(12)__ section, but if the heart is cut so that anterior and posterior parts result, the section is a __(13)__ section. You are told to cut an animal along two planes so that the paired kidneys are observable in both sections. The two sections that meet this requirement are the __(14)__ and __(15)__ sections.

**18.** Using key choices, identify the body cavities where the following body organs are located. Enter the appropriate letter or term in the answer blanks.

### Key Choices

A. Abdominopelvic    B. Cranial    C. Spinal    D. Thoracic

_____ 1. Stomach

_____ 2. Small intestine

_____ 3. Large intestine

_____ 4. Spleen

_____ 5. Liver

_____ 6. Spinal cord

_____ 7. Bladder

_____ 8. Trachea

_____ 9. Lungs

_____ 10. Pituitary gland

_____ 11. Rectum

_____ 12. Ovaries

**19.** Refer to the organs listed in Exercise 18. In the spaces provided, record the numbers of the organs that would be found in each of the abdominal regions named here. Some organs may be found in more than one abdominal region.

_____ 1. Hypogastric region

_____ 2. Right lumbar region

_____ 3. Umbilical region

_____ 4. Epigastric region

_____ 5. Left iliac region

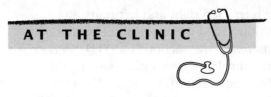

**20.** A jogger has stepped in a pothole and sprained his ankle. What systems have suffered damage?

**21.** A newborn baby is unable to hold down any milk. Examination reveals a developmental disorder in which the esophagus fails to connect to the stomach. What survival needs are most immediately threatened?

**22.** The Chan family was traveling in their van and had a minor accident. The children in the back seat were wearing lap belts but they still sustained bruises around the abdomen and had some internal organ injuries. Why is this area more vulnerable to damage than others?

**23.** John, a patient at Jones City Hospital, is in tough shape. He has a hernia in his inguinal region, pain from an infected kidney in his lumbar region, and severe bruises and swelling in his pubic region. Explain where each of these regions is located.

# Basic Chemistry

Everything in the universe is composed of one or more elements, the unique building blocks of all matter. Although over 100 elemental substances exist, only four of these (carbon, hydrogen, oxygen, and nitrogen) make up over 96% of all living material.

The student activities in this chapter consider basic concepts of both inorganic and organic chemistry. Chemistry is the science that studies the composition of matter. Inorganic chemistry studies the chemical composition of nonliving substances that (generally) do not contain carbon. Organic chemistry studies the carbon-based chemistry (or biochemistry) of living organisms, whether they are maple trees, fish, or humans.

Understanding atomic structure, bonding behavior of elements, and the structure and activities of the most abundant biologic molecules (proteins, fats, carbohydrates, and nucleic acids) is tested in various ways. Mastering these concepts is necessary to understand how the body functions.

## CONCEPTS OF MATTER AND ENERGY

**1.** Select *all* phrases that apply to each of the following statements and insert the letters in the answer blanks.

_____ 1. The energy located in the bonds of food molecules:
   A. is called thermal energy
   B. is a form of potential energy
   C. causes molecular movement
   D. can be transformed to the bonds of ATP

_____ 2. Heat is:
   A. thermal energy
   B. infrared radiation
   C. kinetic energy
   D. molecular movement

_____ 3. Whenever energy is transformed:
   A. the amount of useful energy decreases
   B. some energy is lost as heat
   C. some energy is created
   D. some energy is destroyed

**2.** Use choices from the key to identify the energy *form* in use in each of the following examples.

*Key Choices*

A. Chemical          B. Electrical          C. Mechanical          D. Radiant

_____ 1. Chewing food

_____ 2. Vision (two types, please—think!)

_____ 3. Bending your fingers to make a fist

_____ 4. Breaking the bonds of ATP molecules to energize your muscle cells to make that fist

_____ 5. Lying under a sunlamp

# COMPOSITION OF MATTER

**3.** Complete the following table by inserting the missing words.

| Particle | Location | Electrical charge | Mass |
|---|---|---|---|
|  |  | + 1 |  |
| Neutron |  |  |  |
|  | Orbitals |  |  |

**4.** Insert the *chemical symbol* (the chemist's shorthand) in the answer blank for each of the following elements.

_____ 1. Oxygen      _____ 4. Iodine      _____ 7. Calcium      _____ 10. Magnesium

_____ 2. Carbon      _____ 5. Hydrogen      _____ 8. Sodium      _____ 11. Chloride

_____ 3. Potassium      _____ 6. Nitrogen      _____ 9. Phosphorus      _____ 12. Iron

**5.** Using the key choices, select the correct responses to the following descriptive statements. Insert the appropriate answers in the answer blanks.

*Key Choices*

A. Atom          C. Element          E. Ion          G. Molecule          I. Protons

B. Electrons          D. Energy          F. Matter          H. Neutrons          J. Valence

_____ 1. An electrically charged atom or group of atoms

_____ 2. Anything that takes up space and has mass (weight)

_____ 3. A unique substance composed of atoms having the same atomic number

_____ 4. Negatively charged particles, forming part of an atom

_____ 5. Subatomic particles that determine an atom's chemical behavior, or bonding ability

_____ 6. The ability to do work

_____ 7. The smallest particle of an element that retains the properties of the element

_____ 8. The smallest particle of a compound, formed when atoms combine chemically

_____ 9. Positively charged particles forming part of an atom

_____ 10. Name given to the electron shell that contains the most reactive electrons

_____ 11. _____ 12. Subatomic particles responsible for most of an atom's mass

**6.** For each of the following statements that is true, insert *T* in the answer blank. If any of the statements are false, correct the <u>underlined</u> term by inserting your correction in the answer blank.

_____ 1. $Na^+$ and $K^+$ are <u>needed</u> for nerve cells to conduct electrical impulses.

_____ 2. The atomic number of oxygen is 8. Therefore, oxygen atoms always contain 8 <u>neutrons</u>.

_____ 3. The greater the distance of an electron from the nucleus, the <u>less</u> energy it has.

_____ 4. Electrons are located in more or less designated areas of space around the nucleus called <u>orbitals</u>.

_____ 5. An unstable atom that decomposes and emits energy is called <u>retroactive</u>.

_____ 6. <u>Iron</u> is necessary for oxygen transport in red blood cells.

_____ 7. The most abundant negative ion in extracellular fluid is <u>calcium</u>.

_____ 8. The element essential for the production of thyroid hormones is <u>magnesium</u>.

_____ 9. <u>Calcium</u> is found as a salt in bones and teeth.

# MOLECULES, CHEMICAL BONDS, AND CHEMICAL REACTIONS

**7.** Match the terms in Column B to the chemical equations listed in Column A. Enter the correct letter or term in the answer blanks.

|                | Column A | Column B |
|----------------|----------|----------|
| _____ | 1. A + B ⟶ AB | A. Decomposition |
| _____ | 2. AB + CD ⟶ AD + CB | B. Exchange |
| _____ | 3. XY ⟶ X + Y | C. Synthesis |

**8.** Figure 2–1 is a diagram of an atom. Select two different colors and use them to color the coding circles and corresponding structures on the figure. Complete this exercise by responding to the questions that follow, referring to the atom in this figure. Insert your answers in the answer blanks provided.

◯ Nucleus

◯ Electrons

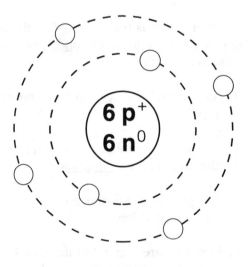

**Figure 2–1**

1. What is the atomic number of this atom? _____

2. What is its atomic mass? _____

3. What atom is this? _____

4. If this atom had one additional neutron but the other subatomic particles remained the same as shown, this slightly different atom (of the same element) would be called a(n) _____

5. Is this atom chemically active or inert? _____

6. How many electrons would be needed to fill its outer (valence) shell? _____

7. Would this atom most likely take part in forming ionic or

   covalent bonds? _____ Why? _____

   _____

9. Both $H_2O_2$ and $2OH^-$ are chemical species with two hydrogen atoms and two oxygen atoms. Briefly explain how these species are different:

   _____

   _____

10. Two types of chemical bonding are shown in Figure 2–2. In the figure, identify each type as a(n) *ionic* or *covalent* bond. In the case of the ionic bond, indicate which atom has lost an electron by adding a colored arrow to show the direction of electron transfer. For the covalent bond, indicate the shared electrons.

**A**

**Type of bond:** _____

**B**

**Type of bond:** _____

**Figure 2–2**

11. Figure 2–3 illustrates five water molecules held together by hydrogen bonds. First, correctly identify the oxygen and hydrogen atoms both by color and by inserting their atomic symbols on the appropriate circles (atoms). Then label the following structures in the figure:

◯ Oxygen

◯ Hydrogen

◯ Positive pole (end)

◯ Negative pole (end)

◯ Hydrogen bonds

**Figure 2–3**

12. Circle each structural formula that is *likely* to be a polar covalent compound.

13. Respond to the instructions following the equation:

$$H_2CO_3 \longrightarrow H^+ + HCO_3^-$$

1. In the space provided, list the chemical formula(s) of compounds. _____

2. In the space provided, list the chemical formula(s) of ions. _____

3. Circle the product(s) of the reaction.

4. Modify the equation by adding a colored arrow in the proper place to indicate that the reaction is reversible.

# BIOCHEMISTRY: THE COMPOSITION OF LIVING MATTER

**14.** Use key choices to identify the substances described in the following statements. Insert the appropriate letter(s) or corresponding term(s) in the answer blanks.

*Key Choices*

A. Acid(s)    B. Base(s)    C. Buffer    D. Salt(s)

_____ 1. _____ 2. _____ 3. Substances that ionize in water; good electrolytes

_____ 4. Proton ($H^+$) acceptor

_____ 5. Ionize in water to release hydrogen ions and a negative ion other than hydroxide ($OH^-$)

_____ 6. Ionize in water to release ions other than $H^+$ and $OH^-$

_____ 7. Formed when an acid and a base are combined

_____ 8. Substances such as lemon juice and vinegar

_____ 9. Prevents rapid/large swings in pH

**15.** Complete the following statements concerning the properties and biological importance of water.

_____ 1.

_____ 2.

_____ 3.

_____ 4.

_____ 5.

_____ 6.

_____ 7.

_____ 8.

The ability of water to maintain a relatively constant temperature and thus prevent sudden changes is because of its high __(1)__. Biochemical reactions in the body must occur in __(2)__. About __(3)__ % of the volume of a living cell is water. Water molecules are bonded to other water molecules because of the presence of __(4)__ bonds. Water, as $H^+$ and $OH^-$ ions, is essential in biochemical reactions such as __(5)__ and __(6)__ reactions. Because of its __(7)__, water is an excellent solvent and forms the basis of mucus and other body __(8)__.

**16.** Using the key choices, fully characterize weak and strong acids.

*Key Choices*

A. Ionize completely in water             E. Ionize at high pH

B. Ionize incompletely in water           F. Ionize at low pH

C. Act as part of a buffer system         G. Ionize at pH 7

D. When placed in water, always act to change the pH

Weak acid: _____      Strong acid: _____

**17.** Use an *X* to designate which of the following are organic compounds.

_____ Carbon dioxide    _____ Fats       _____ Proteins     _____ $H_2O$

_____ Oxygen            _____ KCl        _____ Glucose      _____ DNA

**18.** Match the terms in Column B to the descriptions provided in Column A.
Enter the correct letter(s) or term(s) in the answer blanks.

**Column A**

_____ 1. Building blocks of carbohydrates

_____ 2. Building blocks of fat

_____ 3. Building blocks of protein

_____ 4. Building blocks of nucleic acids

_____ 5. Cellular cytoplasm is primarily composed of this substance

_____ 6. The single most important fuel source for body cells

_____ 7. Not soluble in water

_____ 8. Contain C, H, and O in the ratio $CH_2O$

_____ 9. Contain C, H, and O, but have relatively small amounts of oxygen

_____ 10. _____ 11. These building blocks contain N in addition to C, H, and O

_____ 12. Contain P in addition to C, H, O, and N

_____ 13. Used to insulate the body and found in all cell membranes

_____ 14. Primary components of meat and cheese

_____ 15. Primary components of bread and lollipops

_____ 16. Primary components of egg yolk and peanut oil

**Column B**

A. Amino acids

B. Carbohydrates

C. Lipids (fats)

D. Fatty acids

E. Glycerol

F. Nucleotides

G. Monosaccharides

H. Proteins

_____ 17. Include collagen and hemoglobin

_____ 18. Class that usually includes cholesterol

**19.** Using key choices, correctly select *all* terms that correspond to the following descriptions. Insert the correct letter(s) or their corresponding term(s) in the answer blanks.

**Key Choices**

A. Cholesterol      D. Enzyme        G. Hormones       J. Maltose

B. Collagen         E. Glycogen      H. Keratin        K. RNA

C. DNA              F. Hemoglobin    I. Lactose        L. Starch

_____ 1. Example(s) of fibrous (structural) proteins

_____ 2. Example(s) of globular (functional) proteins

_____ 3. Biologic catalyst

_____ 4. Plant storage carbohydrate

_____ 5. Animal storage carbohydrate

_____ 6. The "stuff" of the genes

_____ 7. A steroid

_____ 8. Double sugars, or disaccharides

**20.** Five simplified diagrams of biological molecules are depicted in Figure 2–4. First, identify the molecules and insert the correct names in the answer blanks on the figure. Then select a different color for each molecule listed below and use them to color the coding circles and the corresponding molecules on the illustration.

◯ Fat                ◯ Nucleotide        ◯ Monosaccharide

◯ Functional protein    ◯ Polysaccharide

A _____    B _____    C _____

D _____

E _____

**Figure 2–4**

**21.** Circle the term that does not belong in each of the following groupings.

    1.  Adenine          Guanine          Glucose          Thymine

    2.  DNA          Ribose          Phosphate          Deoxyribose

    3.  Galactose          Glycogen          Fructose          Glucose

    4.  Amino acid          Polypeptide          Glycerol          Protein

    5.  Glucose          Sucrose          Lactose          Maltose

**22.** For each true statement, insert *T* in the answer blank. If any are false, correct the <u>underlined</u> term and insert your correction in the answer blank.

_____ 1. Phospholipids are <u>polarized</u> molecules.

_____ 2. <u>Steroids</u> are the major form in which body fat is stored.

_____ 3. <u>Water</u> is the most abundant compound in the body.

_____ 4. <u>Nonpolar</u> molecules are generally soluble in water.

_____ 5. The bases of RNA are A, G, C, and <u>U</u>.

_____ 6. The universal energy currency of living cells is <u>RNA</u>.

_____ 7. RNA is <u>single</u> stranded.

_____ 8. The four elements comprising over 90% of living matter are C, H, N, and <u>Na</u>.

**23.** Figure 2–5 shows the molecular structure of DNA, a nucleic acid.

    A. First, identify the two unnamed nitrogen bases and insert their correct names and symbols in the two blanks beside the color-coding circles.

    B. Complete the identification of the bases on the diagram by inserting the correct symbols in the appropriate spaces on the right side of the diagram.

    C. Select different colors and color the coding circles and the corresponding parts of the diagram.

    D. Label one d-R sugar unit and one P unit of the "backbones" of the DNA structure by inserting leader lines and labels on the diagram.

    E. Circle the associated nucleotide.

    ◯ Deoxyribose sugar (d-R)    ◯ Adenine (A)    ◯ _____ ( )

    ◯ Phosphate (P)    ◯ Cytosine (C)    ◯ _____ ( )

Then answer the questions following Figure 2–5 by writing your answers in the answer blanks.

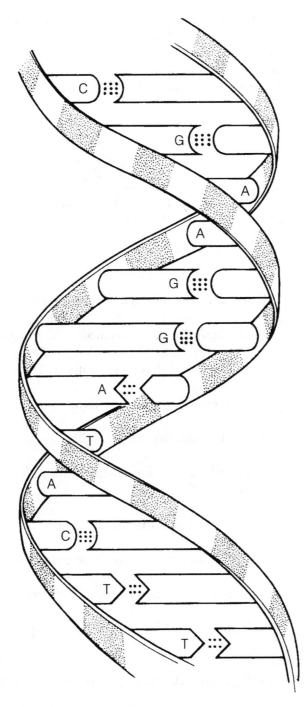

**Figure 2–5**

1. Name the bonds that help to hold the two DNA strands together. _____

2. Name the three-dimensional shape of the DNA molecule. _____

3. How many base-pairs are present in this segment of a DNA model? _____

4. What is the term that means "base-pairing"? _____

**24.** The biochemical reaction shown in Figure 2–6 represents the complete digestion of a polymer (a large molecule as consumed in food) down to its constituent monomers, or building blocks. Select two colors and color the coding circles and the structures. Then, select the one correct answer for each statement below and insert your answer in the answer blank.

○ Monomer          ○ Polymer

**Figure 2–6**

_____  1. If starch is the polymer, the monomer is:

   A. glycogon      B. amino acid      C. glucose      D. maltose

_____  2. During polymer digestion, water as $H^+$ and $OH^-$ ions would:

   A. be a product of the reaction.

   B. act as a catalyst.

   C. enter between monomers, bond to them, and keep them separated.

   D. not be involved in this reaction.

_____  3. Another name for the chemical digestion of polymers is:

   A. dehydration   B. hydrolysis   C. synthesis   D. displacement

_____  4. If the monomers are amino acids, they may differ from each other by their:

   A. R group    B. amino group    C. acid group    D. peptide bond

## INCREDIBLE JOURNEY

### *A Visualization Exercise for Biochemistry*

*. . . you are suddenly up-ended and are carried along
in a sea of water molecules at almost unbelievable speed.*

**25.** Complete the narrative by inserting the missing words in the
answer blanks.

_____  1.

_____  2.

_____  3.

_____  4.

_____  5.

_____  6.

_____  7.

_____  8.

_____  9.

_____  10.

_____  11.

_____  12.

For this journey, you are miniaturized to the size of a very
small molecule by colleagues who will remain in contact with
you by radio. Your instructions are to play the role of a water
molecule and to record any reactions that involve water molecules. Since water molecules are polar molecules, you are
outfitted with an insulated rubber wet suit with one __(1)__
charge at your helmet and two __(2)__ charges, one at the end
of each leg.

As soon as you are injected into your host's bloodstream, you
feel as though you are being pulled apart. Some large, attrac-
tive forces are pulling at your legs from different directions!
You look about but can see only water molecules. After a
moment's thought, you remember the polar nature of your
wet suit. You record that these forces must be the __(3)__ that
are easily formed and easily broken in water.

After this initial surprise, you are suddenly up-ended and carried along in a sea of water molecules at almost unbelievable
speed. You have just begun to observe some huge, red, disk-
shaped structures (probably __(4)__) taking up $O_2$ molecules,
when you are swept into a very turbulent environment. Your
colleagues radio that you are in the small intestine. With difficulty, because of numerous collisions with other molecules,
you begin to record the various types of molecules you see.

In particular, you notice a very long helical molecule made of units with distinctive R-groups. You
identify and record this type of molecule as a __(5)__, made of units called __(6)__ that are joined
together by __(7)__ bonds. As you move too close to the helix during your observations, you are
nearly pulled apart to form two ions, __(8)__, but you breathe a sigh of relief as two ions of another
water molecule take your place. You watch as these two ions move between two units of the long
helical molecule. Then, in a fraction of a second, the bond between the two units is broken. As
you record the occurrence of this chemical reaction, called __(9)__, you are jolted into another direction by an enormous globular protein, the very same __(10)__ that controls and speeds up this chemical reaction.

Once again you find yourself in the bloodstream, heading into an organ identified by your colleagues as the liver. Inside a liver cell, you observe many small monomers, made up only of C, H,
and O atoms. You identify these units as __(11)__ molecules because the liver cells are bonding them
together to form very long, branched polymers called __(12)__. You record that this type of chemical

_____ 13.

_____ 14.

_____ 15.

reaction is called __(13)__, and you happily note that this reaction also produces __(14)__ molecules like you!

After another speedy journey through the bloodstream, you reach the skin. You move deep into the skin and finally gain access to a sweat gland. In the sweat gland, you collide with millions of water molecules and some ionized salt molecules that are continually attracted to your positive and negative charges. Suddenly, the internal temperature rises, and molecular collisions __(15)__ at an alarming rate, propelling you through the pore of the sweat gland onto the surface of the skin. So that you will be saved from the fate of evaporating into thin air, you contact your colleagues and are speedily rescued.

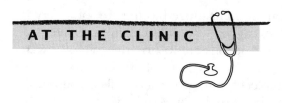

## AT THE CLINIC

26. It is determined that a patient is in acidosis. What does this mean, and would you treat the condition with a chemical that would *raise* or *lower* the pH?

27. A newborn is diagnosed with sickle-cell anemia, a genetic disease in which substitution of one amino acid results in abnormal hemoglobin. Explain to the parents how the substitution can have such a drastic effect on the structure of the protein.

28. Johnny's body temperature is spiking upward. When it reaches 104°F, his mother puts in a call to the pediatrician. She is advised to give Johnny children's aspirin and sponge his body with cool to tepid water to prevent a further rise in temperature. How might a fever (excessively high body temperature) be detrimental to Johnny's welfare?

29. Mrs. Gallo's physician suspects that she is showing the initial signs of multiple sclerosis, a disease characterized by the formation of hardened plaques in the insulating sheaths surrounding nerve fibers. What medical imaging technique will the physician probably order to determine if such plaques are present?

# 3 Cells and Tissues

The basic unit of structure and function in the human body is the cell. Each of a cell's parts, or organelles, as well as the entire cell, is organized to perform a specific function. Cells have the ability to metabolize, grow and reproduce, move, and respond to stimuli. The cells of the body differ in shape, size, and in specific roles in the body. Cells that are similar in structure and function form tissues, which, in turn, construct the various body organs.

Student activities in this chapter include questions relating to the structure and function of the generalized animal cell and to the general arrangement of tissues and their contribution to the activities of the various body organs.

## CELLS

### Overview

**1.** Answer the following questions by inserting your responses in the answer blanks.

_____ 1.

_____ 2.

_____ 3.

_____ 4.

_____ 5.

_____ 6.

_____ 7.

_____ 8.

_____ 9.

_____ 10.

1–4. Name the four elements that make up the bulk of living matter.

5. Name the single most abundant material or substance in living matter.

6. Name the trace element most important for making bones hard.

7. Name the element, found in small amounts in the body, that is needed to make hemoglobin for oxygen transport.

8–12. Although there are many specific "jobs" that certain cells are able to do, name five functions common to all cells.

_____ 11.

_____ 12.   →

_____ 13.    13–15. List three different cell shapes.

_____ 14.    16. Name the fluid, similar to seawater, that surrounds and bathes all body cells.

_____ 15.

_____ 16.    17. Name the flattened cells, important in protection, that fit together like tiles. (This is just one example of the generalization that a cell's structure is very closely related to its function in the body.)

_____ 17.

## Anatomy of a Generalized Cell

2. Complete the following table to fully describe the various cell parts. Insert your responses in the spaces provided under each heading.

| Cell structure | Location | Function |
|---|---|---|
| | External boundary of the cell | Confines cell contents; regulates entry and exit of materials |
| Lysosomes | | |
| | Scattered throughout the cell | Control release of energy from foods; form ATP |
| | Projections of the plasma membrane | Increase the membrane surface area |
| Golgi apparatus | | |
| Nucleus | | |
| | Two rod-shaped bodies near the nucleus | Direct formation of the mitotic spindle |
| Nucleolus | | |
| Smooth ER | | |
| Rough ER | | |
| | Attached to membrane systems or scattered in the cytoplasm | Synthesize proteins |
| Chromatin | | |
| | Scattered in cytoplasm | Detoxify alcohol, hydrogen peroxide, etc. |
| Inclusions | | |

**3.** Using the following list of terms, correctly label all cell parts indicated by
leader lines in Figure 3–1. Then select different colors for each structure
and use them to color the coding circles and the corresponding structures
in the illustration.

◯ Plasma membrane        ◯ Mitochondrion

◯ Centriole(s)        ◯ Nuclear membrane

◯ Chromatin thread(s)        ◯ Nucleolus

◯ Golgi apparatus        ◯ Rough endoplasmic reticulum (ER)

◯ Microvilli        ◯ Smooth endoplasmic reticulum (ER)

Cytosol

Vacuole

**Figure 3–1**

# Cell Physiology

## Membrane Transport

**4.** A semipermeable sac, containing 4% NaCl, 9% glucose, and 10% albumin, is suspended in a solution with the following composition: 10% NaCl, 10% glucose, and 40% albumin. Assume the sac is permeable to all substances *except* albumin. Using the key choices, insert the letter indicating the correct event in the answer blanks.

### Key Choices

A. Moves into the sac          B. Moves out of the sac          C. Does not move

_____ 1. Glucose          _____ 3. Albumin

_____ 2. Water          _____ 4. NaCl

**5.** Figure 3–2 shows three microscopic fields (A–C) containing red blood cells. Arrows indicate the direction of net osmosis. Respond to the following questions, referring to Figure 3–2, by inserting your responses in the spaces provided.

1. Which microscopic field contains a *hypertonic* solution? _____

   The cells in this field are said to be _____

2. Which microscopic field contains an isotonic bathing solution? _____

   What does *isotonic* mean? _____

   _____

3. Which microscopic field contains a *hypotonic* solution? _____

   What is happening to the cells in this field and why? _____

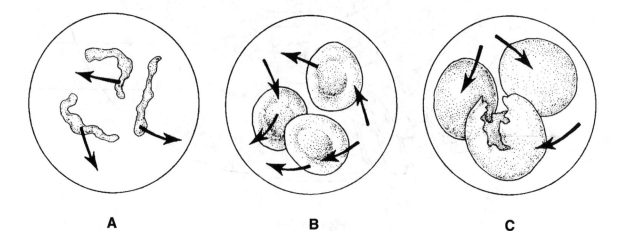

A                              B                              C

**Figure 3–2**

**6.** Select the key choices that characterize each of the following statements. Insert the appropriate letter(s) or corresponding term(s) in the answer blanks.

*Key Choices*

A. Diffusion, simple        C. Endocytosis, phagocytosis        E. Filtration

B. Diffusion, osmosis       D. Exocytosis                       F. Solute pumping

_____ 1. Require ATP (cellular energy)

_____ 2. Driven by kinetic energy of the molecules

_____ 3. Driven by hydrostatic (fluid) pressure

_____ 4. Follow a concentration gradient

_____ 5. Proceeds against a concentration gradient; require(s) a carrier

_____ 6. A means of secreting cell products

_____ 7. Moves water through a semipermeable membrane

_____ 8. Transports amino acids, some sugars, and Na$^+$ through the plasma membrane

_____ 9. Provides for cellular uptake of solid or large particles from the cell exterior

_____ 10. Moves small or lipid-soluble solutes through the membrane

**7.** Figure 3–3 represents a portion of a plasma membrane. Select two different colors for lipid and protein molecules. Color the coding circles and the corresponding molecules in the illustration. Then add a colored arrow for each substance shown inside and outside the cell indicating (a) its *direction* of transport through the membrane; and (b) its *means of transport* (that is, either directly through the lipid portion or by attachment to a protein carrier).

◯ Lipid molecules

◯ Protein molecules

Two types of molecules not shown here that contribute to plasma membrane structure

are _____ and _____ .

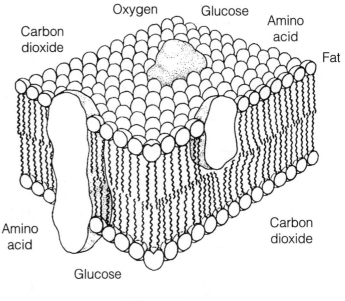

**Cell exterior**

**Cell interior**

**Figure 3–3**

## Cell Division

**8.** The following statements provide an overview of the structure of DNA (genetic material) and its role in the body. Choose responses from the key choices that complete the statements. Insert the appropriate answers in the answer blanks.

### Key Choices

| | | | |
|---|---|---|---|
| A. Adenine | G. Enzymes | M. Nucleotides | S. Ribosome |
| B. Amino acids | H. Genes | N. Old | T. Sugar (deoxyribose) |
| C. Bases | I. Growth | O. Phosphate | U. Template, or model |
| D. Codons | J. Guanine | P. Proteins | V. Thymine |
| E. Complementary | K. Helix | Q. Replication | W. Transcription |
| F. Cytosine | L. New | R. Repair | X. Uracil |

_____ 1.

_____ 2.

_____ 3.

_____ 4.

_____ 5.

_____ 6.

_____ 7.

_____ 8.

_____ 9.

_____ 10.

_____ 11.

_____ 12.

_____ 13.

_____ 14.

_____ 15.

_____ 16.

_____ 17.

_____ 18.

DNA molecules contain information for building specific __(1)__ . In a three-dimensional view, a DNA molecule looks like a spiral staircase; this is correctly called a __(2)__ . The constant parts of DNA molecules are the __(3)__ and __(4)__ molecules, forming the DNA-ladder uprights, or backbones. The information of DNA is actually coded in the sequence of nitrogen-containing __(5)__ , which are bound together to form the "rungs" of the DNA ladder. When the four DNA bases are combined in different three-base sequences, called triplets, different __(6)__ of the protein are called for. It is said that the N-containing bases of DNA are __(7)__ , which means that only certain bases can fit or interact together. Specifically, this means that __(8)__ can bind with guanine, and adenine binds with __(9)__ .

The production of proteins involves the cooperation of DNA and RNA. RNA is another type of nucleic acid that serves as a "molecular slave" to DNA. That is, it leaves the nucleus and carries out the instructions of the DNA for the building of a protein on a cytoplasmic structure called a __(10)__ . When a cell is preparing to divide, in order for its daughter cells to have all its information, it must oversee the __(11)__ of its DNA so that a "double dose" of genes is present for a brief period. For DNA synthesis to occur, the DNA must uncoil, and the bonds between the N-bases must be broken. Then the two single strands of __(12)__ each act as a __(13)__ for the building of a whole DNA molecule. When completed, each DNA molecule formed is half __(14)__ and half __(15)__ . The fact that DNA replicates before a cell divides ensures that each daughter cell has a complete set of __(16)__ . Cell division, which then follows, provides new cells so that __(17)__ and __(18)__ can occur.

**9.** Identify the phases of mitosis depicted in Figure 3–4 by inserting the correct name in the blank under the appropriate diagram. Then select different colors to represent the structures listed below and use them to color in the coding circles and the corresponding structures in the illustration.

◯ Nuclear membrane(s), if present      ◯ Centrioles

◯ Nucleoli, if present      ◯ Spindle fibers

◯ Chromosomes

A _____

B _____

C _____

D _____

**Figure 3–4**

**10.** The following statements describe events that occur during the different phases of mitosis. Identify the phase by choosing the correct response(s) from key choices and inserting the letter(s) or term(s) in the answer blanks.

*Key Choices*

A. Anaphase          C. Prophase          E. None of these

B. Metaphase         D. Telophase

_____ 1. Chromatin coils and condenses to form deeply staining bodies.

_____ 2. Centromeres break, and chromosomes begin migration toward opposite poles of the cell.

_____ 3. The nuclear membrane and nucleoli reappear.

_____ 4. When chromosomes cease their poleward movement, this phase begins.

_____ 5. Chromosomes align on the equator of the spindle.

_____ 6. The nucleoli and nuclear membrane disappear.

_____ 7. The spindle forms through the migration of the centrioles.

_____ 8. Chromosomal material replicates.

_____ 9. Chromosomes first appear to be duplex structures.

_____ 10. Chromosomes attach to the spindle fibers.

_____ 11. A cleavage furrow forms during this phase.

_____ 12. The nuclear membrane is absent during the entire phase.

_____ 13. Period during which a cell carries out its *usual* metabolic activities.

**11.** Complete the following statements. Insert your answers in the answer blanks.

_____ 1.

_____ 2.

_____ 3.

_____ 4.

_____ 5.

_____ 6.

_____ 7.

Division of the __(1)__ is referred to as mitosis. Cytokinesis is division of the __(2)__. The major structural difference between chromatin and chromosomes is that the latter are __(3)__. Chromosomes attach to the spindle fibers by undivided structures called __(4)__. If a cell undergoes nuclear division but not cytoplasmic division, the product is a __(5)__. The structure that acts as a scaffolding for chromosomal attachment and movement is called the __(6)__. __(7)__ is the period of cell life when the cell is not involved in division.

*Protein Synthesis*

**12.** Figure 3–5 is a diagram illustrating protein synthesis. Select four different colors, and use them to color the coding circles and the corresponding structures in the diagram. Next, using the letters of the genetic code, label the nitrogen bases on strand 2 of the DNA double helix, on the mRNA strands, and on the tRNA molecules. Then, answer the questions that follow referring to Figure 3–5, inserting your answers in the answer blanks.

◯ Backbones of the DNA double helix       ◯ tRNA molecules

◯ Backbone of the mRNA strands           ◯ Amino acid molecules

**Figure 3–5**

1. Transfer of the genetic message from DNA to mRNA is called _____.

2. Assembly of amino acids according to the genetic information carried by mRNA is called

    _____.

3. The set of three nitrogen bases on tRNA that is complementary to an mRNA codon is called

    a _____. The complementary three-base sequence on DNA is called a

    _____.

# BODY TISSUES

**13.** Twelve tissue types are diagrammed in Figure 3–6. Identify each tissue type by inserting the correct name in the blank below it on the diagram. Select different colors for the following structures and use them to color the coding circles and corresponding structures in the diagrams.

◯ Epithelial cells

◯ Muscle cells

◯ Nerve cells

◯ Matrix (Where found, matrix should be colored differently from the living cells of that tissue type. Be careful, this may not be as easy as it seems!)

A _____

B _____

C _____

D _____

E _____

F _____

**Figure 3–6, A–F**

G _____  H _____

I _____  J _____

K _____  L _____

**Figure 3-6, G-L**

**14.** Describe briefly how the particular structure of a neuron relates to its function

in the body. _____

_____

_____

**15.** Using key choices, correctly identify the *major* tissue types described. Enter the appropriate letter or tissue type term in the answer blanks.

*Key Choices*

A. Connective      B. Epithelium      C. Muscle      D. Nervous

_____ 1. Forms mucous, serous, and epidermal membranes

_____ 2. Allows for organ movements within the body

_____ 3. Transmits electrochemical impulses

_____ 4. Supports body organs

_____ 5. Cells of this tissue may absorb and/or secrete substances

_____ 6. Basis of the major controlling system of the body

_____ 7. The cells of this tissue shorten to exert force

_____ 8. Forms hormones

_____ 9. Packages and protects body organs

_____ 10. Characterized by having large amounts of nonliving matrix

_____ 11. Allows you to smile, grasp, swim, ski, and shoot an arrow

_____ 12. Most widely distributed tissue type in the body

_____ 13. Forms the brain and spinal cord

**16.** Using key choices, identify the following specific type(s) of epithelial tissue. Enter the appropriate letter or classification term in the answer blanks.

*Key Choices*

A. Pseudostratified columnar (ciliated)      C. Simple cuboidal      E. Stratified squamous

B. Simple columnar      D. Simple squamous      F. Transitional

_____ 1. Lines the esophagus and forms the skin epidermis

_____ 2. Forms the lining of the stomach and small intestine

_____ 3. Best suited for areas subjected to friction

_____ 4. Lines much of the respiratory tract

_____ 5. Propels substances (e.g., mucus) across its surface

_____ 6. Found in the bladder lining; peculiar cells that slide over one another

_____ 7. Forms thin serous membranes; a single layer of flattened cells

**17.** Epithelium exhibits many plasma membrane modifications. Figure 3–7 depicts some of these modifications.

First: Choose a color for the coding circles and the corresponding structures in the figure.

○ Epithelial cell cytoplasm          ○ Connective tissue

○ Epithelial cell nucleus          ○ Blood vessel

○ Nerve fibers

Second: Correctly identify the following structures or regions by labeling the appropriate leader lines using terms from the list below:

A. Apical region          D. Cilia          G. Epithelium

B. Basement membrane          E. Connective tissue          H. Microvilli

C. Capillary          F. Desmosome          I. Tight junctions

**Figure 3–7**

**18.** The three types of muscle tissue exhibit certain similarities and differences. Check (✓) the appropriate spaces in the following table to indicate which muscle types exhibit each characteristic.

| Characteristic | Skeletal | Cardiac | Smooth |
|---|---|---|---|
| 1. Voluntarily controlled | | | |
| 2. Involuntarily controlled | | | |
| 3. Banded appearance | | | |
| 4. Single nucleus in each cell | | | |
| 5. Multinucleate | | | |
| 6. Found attached to bones | | | |
| 7. Allows you to direct your eyeballs | | | |
| 8. Found in the walls of stomach, uterus, and arteries | | | |
| 9. Contains spindle-shaped cells | | | |
| 10. Contains cylindrical cells with branching ends | | | |
| 11. Contains long, nonbranching cylindrical cells | | | |
| 12. Displays intercalated disks | | | |
| 13. Concerned with locomotion of the body as a whole | | | |
| 14. Changes the internal volume of an organ as it contracts | | | |
| 15. Tissue of the circulatory pump | | | |

**19.** Circle the term that does not belong in each of the following groupings.

1. Collagen      Cell      Matrix      Cell product

2. Cilia      Flagellum      Microvilli      Elastic fibers

3. Glands      Bones      Epidermis      Mucosae

4. Adipose      Hyaline      Osseous      Nervous

5. Blood      Smooth      Cardiac      Skeletal

**20.** Using key choices, identify the following connective tissue types. Insert the appropriate letter or corresponding term in the answer blanks.

*Key Choices*

A. Adipose connective tissue      C. Dense fibrous connective tissue      E. Reticular connective tissue

B. Areolar connective tissue       D. Osseous tissue                               F. Hyaline cartilage

_____ 1. Provides great strength through parallel bundles of collagenic fibers; found in tendons

_____ 2. Acts as a storage depot for fat

_____ 3. Composes the dermis of the skin

_____ 4. Forms the bony skeleton

_____ 5. Composes the basement membrane and packages organs; includes a gel-like matrix with all categories of fibers and many cell types

_____ 6. Forms the embryonic skeleton and the surfaces of bones at the joints; reinforces the trachea

_____ 7. Provides insulation for the body

_____ 8. Structurally amorphous matrix, heavily invaded with fibers; appears glassy and smooth

_____ 9. Contains cells arranged concentrically around a nutrient canal; matrix is hard due to calcium salts

_____ 10. Forms the stroma or internal "skeleton" of lymph nodes, the spleen, and other lymphoid organs

# Tissue Repair

**21.** For each of the following statements about tissue repair that is true, enter *T* in the answer blank. For each false statement, correct the underlined words by writing the correct words in the answer blank.

_____ 1. The nonspecific response of the body to injury is called <u>regeneration</u>.

_____ 2. Intact capillaries near an injury dilate, leaking plasma, blood cells, and <u>antibodies</u>, which cause the blood to clot. The clot at the surface dries to form a scab.

_____ 3. During the first phase of tissue repair, capillary buds invade the clot, forming a delicate pink tissue called <u>endodermal</u> tissue.

_____ 4. When damage is not too severe, the surface epithelium migrates beneath the dry scab and across the surface of the granulation tissue. This repair process is called <u>proliferation</u>.

_____ 5. If tissue damage is very severe, tissue repair is more likely to occur by <u>fibrosis</u>, or scarring.

_____ 6. During fibrosis, fibroblasts in the granulation tissue lay down <u>keratin</u> fibers, which form a strong, compact, but inflexible mass.

_____ 7. The repair of cardiac muscle and nervous tissue occurs only by <u>fibrosis</u>.

# DEVELOPMENTAL ASPECTS OF CELLS AND TISSUES

**22.** Correctly complete each statement by inserting your responses in the answer blanks.

_____ 1.

_____ 2.

_____ 3.

_____ 4.

_____ 5.

_____ 6.

_____ 7.

_____ 8.

_____ 9.

_____ 10.

_____ 11.

_____ 12.

_____ 13.

_____ 14.

_____ 15.

_____ 16.

_____ 17.

_____ 18.

_____ 19.

_____ 20.

During embryonic development, cells specialize to form __(1)__. Mitotic cell division is very important for overall body __(2)__. All tissues except __(3)__ tissue continue to undergo cell division until the end of adolescence. After this time, __(4)__ tissue also becomes amitotic. When amitotic tissues are damaged, they are replaced by __(5)__ tissue, which does not function in the same way as the original tissue. This is a serious problem when heart cells are damaged.

Aging begins almost as soon as we are born. Three explanations of the aging process have been offered. One states that __(6)__ insults, such as the presence of toxic substances in the blood, are important. Another theory states that external __(7)__ factors, such as X-rays, help to cause aging. A third theory suggests that aging is programmed in our __(8)__. Three examples of aging processes seen in all people are __(9)__, __(10)__, and __(11)__.

Neoplasms occur when cells "go wild" and the normal controls of cell __(12)__ are lost. The two types of neoplasms are __(13)__ and __(14)__. The __(15)__ type tends to stay localized and have a capsule. The __(16)__ type is likely to invade other body tissues and spread to other (distant) parts of the body. To correctly diagnose the type of neoplasm, a microscopic examination of the tissue called a __(17)__ is usually done. Whenever possible, __(18)__ is the treatment of choice for neoplasms.

An overgrowth of tissue that is not considered to be a neoplasm is referred to as __(19)__. Conversely, a decrease in the size of an organ or tissue, resulting from loss of normal stimulation, is called __(20)__.

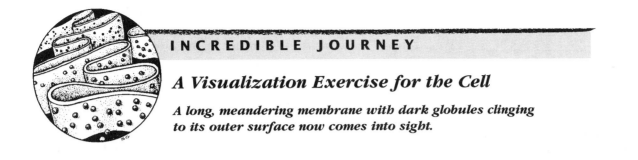

# INCREDIBLE JOURNEY

## A Visualization Exercise for the Cell

*A long, meandering membrane with dark globules clinging to its outer surface now comes into sight.*

**23.** Where necessary, complete statements by inserting the missing words in the answer blanks.

_____ 1.

_____ 2.

_____ 3.

_____ 4.

_____ 5.

_____ 6.

_____ 7.

_____ 8.

_____ 9.

_____ 10.

For your second journey, you will be miniaturized to the size of a small protein molecule and will travel in a microsubmarine, specially designed to enable you to pass easily through living membranes. You are injected into the intercellular space between two epithelial cells, and you are instructed to observe one of these cells firsthand, and to identify as many of its structures as possible.

You struggle briefly with the controls and then maneuver your microsub into one of these cells. Once inside the cell, you find yourself in a kind of "sea." This salty fluid that surrounds you is the __(1)__ of the cell.

Far below looms a large, dark, oval structure, much larger than anything else you can see. You conclude that it is the __(2)__. As you move downward, you pass a cigar-shaped structure with strange-looking folds on its inner surface. Although you have a pretty good idea that it must be a __(3)__, you decide to investigate more thoroughly. After passing through the external membrane of the structure, you are confronted with yet another membrane. Once past this membrane, you are inside the strange-looking structure. You activate the analyzer switch in your microsub for a readout indicating which molecules are in your immediate vicinity. As suspected, there is an abundance of energy-rich __(4)__ molecules. Having satisfied your curiosity, you leave this structure to continue the investigation.

A long, meandering membrane with dark globules clinging to its outer surface now comes into sight. You maneuver closer and sit back to watch the activity. As you watch, amino acids are joined together and a long, threadlike protein molecule is built. The globules must be __(5)__, and the membrane, therefore, is the __(6)__. Once again, you head toward the large dark structure seen and tentatively identified earlier. On approach, you observe that this huge structure has very large openings in its outer wall; these openings must be the __(7)__. Passing through one of these openings, you discover that from the inside the color of this structure is a result of dark, coiled, intertwined masses of __(8)__, which your analyzer confirms contain genetic material, or __(9)__ molecules. Making your way through this tangled mass, you pass two round, dense structures that appear to be full of the same type of globules you saw outside. These two round structures are __(10)__. All this information confirms your earlier identification of this cellular structure, so now you move to its exterior to continue observations.

_____ 11.   Just ahead, you see what appears to be a mountain of flat-
                              tened sacs with hundreds of small sac-like vesicles at its
_____ 12.   edges. The vesicles seem to be migrating away from this area
                              and heading toward the outer edges of the cell. The mountain
                              of sacs must be the  (11) . Eventually you come upon a
rather simple-looking membrane-bound sac. Although it doesn't look too exciting, and has few
distinguishing marks, it does not resemble anything else you have seen so far. Deciding to obtain
a chemical analysis before entering this sac, you activate the analyzer and on the screen you see
"Enzymes — Enzymes — Hydrolases — Hydrolases — Danger — Danger." There is little doubt
that this innocent-appearing structure is actually a  (12) .

Completing your journey, you count the number of organelles identified so far. Satisfied that you
have observed most of them, you request retrieval from the intercellular space.

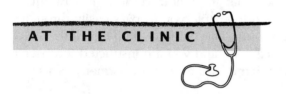

AT THE CLINIC

24. Johnny lacerated his arm and rushed home to Mom so she could "fix it." His
    mother poured hydrogen peroxide over the area and it bubbled vigorously
    where it came in contact with the wound. Since you can expect that cells
    were ruptured in the injured area, what do you _think_ was happening here?

25. The epidermis (epithelium of the cutaneous membrane or skin) is a
    keratinized stratified squamous epithelium. Explain why that epithelium is
    much better suited for protecting the body's external surface than a mucosa
    consisting of a simple columnar epithelium would be.

26. Streptomycin (an antibiotic) binds to the small ribosomal subunit of bacteria
    (but not to the ribosomes of the host cells infected by bacteria). The result is
    the misreading of bacteria mRNA and the breakup of polysomes. What
    process is being affected, and how does this kill the bacterial cells?

**27.** Systemic lupus erythematosus (often simply called lupus) is a condition that sometimes affects young women. It is a chronic (persistent) inflammation that affects all or most of the connective tissue proper in the body. Suzy is told by her doctor that she has lupus, and she asks if it will have widespread or merely localized effects within the body. What would the physician answer?

**28.** Mrs. Linsey sees her gynecologist because she is unable to become pregnant. The doctor discovers granulation tissue in her vaginal canal and explains that sperm are susceptible to some of the same chemicals as bacteria. What is inhibiting the sperm?

# 4

# Skin and Body Membranes

Body membranes, which cover body surfaces, line its cavities, and form protective sheets around organs, fall into two major categories. These are epithelial membranes (skin epidermis, mucosae, and serosae) and the connective tissue synovial membranes.

Topics for review in this chapter include a comparison of structure and function of various membranes, anatomical characteristics of the skin (composed of the connective tissue dermis and the epidermis) and its derivatives, and the manner in which the skin responds to both internal and external stimuli to protect the body.

## CLASSIFICATION OF BODY MEMBRANES

1. Complete the following table relating to body membranes. Enter your responses in the areas left blank.

| Membrane | Tissue type (epithelial/connective) | Common locations | Functions |
|---|---|---|---|
| Mucous | | | |
| Serous | | | |
| Cutaneous | | | |
| Synovial | | | |

2. Four simplified diagrams are shown in Figure 4–1. Select different colors for the membranes listed below, and use them to color the coding circles and the corresponding structures.

○ Cutaneous membrane  ○ Parietal pleura (serosa)  ○ Synovial membrane

○ Mucosae  ○ Visceral pericardium (serosa)

○ Visceral pleura (serosa)  ○ Parietal pericardium (serosa)

**Figure 4–1**

# INTEGUMENTARY SYSTEM (SKIN)
## Basic Structure and Function

**3.** Figure 4–2 depicts a longitudinal section of the skin. Label the skin structures and areas indicated by leader lines and brackets on the figure. Select different colors for the structures below and color the coding circles and the corresponding structures on the figure.

◯ Arrector pili muscle

◯ Adipose tissue

◯ Hair follicle

◯ Nerve fibers

◯ Sweat (sudoriferous) gland

◯ Sebaceous gland

**Figure 4–2**

**4.** The more superficial cells of the epidermis become less viable and ultimately die. What two factors account for this natural demise of the epidermal cells?

1. _____

2. _____

_____

**5.** Complete the following statements in the blanks provided.

_____  1. Radiation from the skin surface and evaporation of sweat are two ways in which the skin helps to get rid of body __(1)__.

_____  2. Fat in the __(2)__ tissue layer beneath the dermis helps to insulate the body.

_____  3. The waterproofing protein found in the epidermal cells is called __(3)__.

_____  4. A vitamin that is manufactured in the skin is __(4)__.

_____  5. A localized concentration of melanin is __(5)__.

_____  6. Wrinkling of the skin is due to loss of the __(6)__ of the skin.

_____  7. A decubitus ulcer results when skin cells are deprived of __(7)__.

_____  8. __(8)__ is a bluish cast of the skin resulting from inadequate oxygenation of the blood.

**6.** Using key choices, choose all responses that apply to the following descriptions. Enter the appropriate letter(s) or term(s) in the answer blanks.

***Key Choices***

A. Stratum corneum          D. Stratum lucidum          G. Epidermis as a whole

B. Stratum basale           E. Papillary layer           H. Dermis as a whole

C. Stratum granulosum       F. Reticular layer

_____  1. Translucent cells, containing keratin

_____  2. Dead cells

_____  3. Dermis layer responsible for fingerprints

_____  4. Vascular region

_____  5. Epidermal region involved in rapid cell division; most inferior epidermal layer

_____  6. Scalelike cells full of keratin that constantly flake off

_____  7. Site of elastic and collagen fibers

_____  8. Site of melanin formation

_____  9. Major skin area from which the derivatives (hair, nails) arise

# Appendages

**7.** For each true statement, write *T*. For each false statement, correct the
underlined word(s) and insert your correction in the answer blank.

_____ 1. Greater amounts of the pigment <u>carotene</u> are produced when
the skin is exposed to the sun.

_____ 2. The most abundant protein in dead epidermal structures such
as hair and nails is <u>melanin</u>.

_____ 3. <u>Sebum</u> is an oily mixture of lipids, cholesterol, and cell fragments.

_____ 4. The oldest epidermal cells in the epidermis are found in the
<u>stratum basale</u>.

_____ 5. The externally observable part of a hair is called the <u>root</u>.

_____ 6. The <u>epidermis</u> provides mechanical strength to the skin.

**8.** Figure 4–3 is a diagram of a cross-sectional view of a hair in its follicle.
Complete this figure by following the directions in steps 1–3.

1. Identify the two portions of the follicle wall by placing the correct name
of the sheath at the end of the appropriate leader line.

2. Use different colors to color these regions.

3. Label, color code, and color the three following regions of the hair.

◯ Cortex          ◯ Cuticle          ◯ Medulla

Hair

Follicle
wall

**Figure 4–3**

**9.** Using key choices, complete the following statements. Insert the appropriate letter(s) or term(s) in the answer blanks.

*Key Choices*

A. Arrector pili        C. Hair            E. Sebaceous glands      G. Sweat gland (eccrine)

B. Cutaneous receptors   D. Hair follicle(s)   F. Sweat gland (apocrine)

_____ 1. A blackhead is an accumulation of oily material produced by \_(1)\_ .

_____ 2. Tiny muscles attached to hair follicles that pull the hair upright during fright or cold are called \_(2)\_ .

_____ 3. The most numerous variety of perspiration gland is the \_(3)\_ .

_____ 4. A sheath formed of both epithelial and connective tissues is the \_(4)\_ .

_____ 5. A less numerous variety of perspiration gland is the \_(5)\_ . Its secretion (often milky in appearance) contains proteins and other substances that favor bacterial growth.

_____ 6. \_(6)\_ is found everywhere on the body except the palms of the hands, soles of the feet, and lips, and primarily consists of dead keratinized cells.

_____ 7. \_(7)\_ are specialized nerve endings that respond to temperature and touch, for example.

_____ 8. \_(8)\_ become more active at puberty.

_____ 9. Part of the heat-liberating apparatus of the body is the \_(9)\_ .

**10.** Circle the term that does not belong in each of the following groupings.

1. Sebaceous gland        Hair            Arrector pili            Epidermis

2. Radiation         Absorption            Conduction                Evaporation

3. Stratum corneum            Nails            Hair            Stratum basale

4. Freckles         Blackheads            Moles            Melanin

5. Scent glands            Eccrine glands            Apocrine glands            Axilla

6. Cyanosis         Erythema            Wrinkles            Pallor

7. Keratin            Carotene            Melanin            Hemoglobin

# Homeostatic Imbalances of the Skin

**11.** Overwhelming infection is one of the most important causes of death in burn patients. What is the other major problem they face, and what are its possible consequences?

_____

_____

**12.** This section reviews the severity of burns. Using the key choices, select the correct burn type for each of the following descriptions. Enter the correct answers in the answer blanks.

*Key Choices*

A. First-degree burn    B. Second-degree burn    C. Third-degree burn

_____ 1. Full-thickness burn; epidermal and dermal layers destroyed; skin is blanched

_____ 2. Blisters form

_____ 3. Epidermal damage, redness, and some pain (usually brief)

_____ 4. Epidermal and some dermal damage; pain; regeneration is possible

_____ 5. Regeneration impossible; requires grafting

_____ 6. Pain is absent because nerve endings in the area are destroyed

**13.** What is the importance of the "rule of nines" in treatment of burn patients?

_____

**14.** Fill in the type of skin cancer which matches each of the following descriptions:

_____ 1. Epithelial cells, not in contact with the basement membrane, develop lesions; metastasizes.

_____ 2. Cells of the lowest level of the epidermis invade the dermis and hypodermis; exposed areas develop ulcer; slow to metastasize.

_____ 3. Rare but often deadly cancer of pigment-producing cells.

**15.** What does ABCD mean in reference to examination of pigmented areas? _____

_____

_____

# DEVELOPMENTAL ASPECTS OF THE SKIN AND BODY MEMBRANES

**16.** Match the choices (letters or terms) in Column B with the appropriate descriptions in Column A.

| Column A | Column B |
|---|---|
| _____ 1. Skin inflammations that increase in frequency with age | A. Acne |
| _____ 2. Cause of graying hair | B. Cold intolerance |
| _____ 3. Small white bumps on the skin of newborn babies, resulting from accumulations of sebaceous gland material | C. Dermatitis |
|  | D. Delayed-action gene |
| _____ 4. Reflects the loss of insulating subcutaneous tissue with age | E. Lanugo |
| _____ 5. A common consequence of accelerated sebaceous gland activity during adolescence | F. Milia |
| _____ 6. Oily substance produced by the fetus's sebaceous glands | G. Vernix caseosa |
| _____ 7. The hairy "cloak" of the fetus |  |

**INCREDIBLE JOURNEY**

### A Visualization Exercise for the Skin

*Your immediate surroundings resemble huge grotesquely twisted vines. . . . you begin to climb upward.*

**17.** Where necessary, complete statements by inserting the missing words in the answer blanks.

_____ 1.

_____ 2.

For this trip, you are miniaturized for injection into your host's skin. Your journey begins when you are deposited in a soft gel-like substance. Your immediate surroundings resemble huge grotesquely twisted vines. But when you peer carefully at the closest "vine," you realize you are actually seeing connective tissue fibers. Although tangled together, most of the fibers are fairly straight and look like strong cables. You identify these as the __(1)__ fibers. Here and there are fibers that resemble coiled springs. These must be the __(2)__ fibers that help to give skin its springiness. At this point,

_____  3.  there is little question that you are in the ___(3)__ region of the skin, particularly since you can also see blood vessels and

_____  4.  nerve fibers around you.

_____  5.  Carefully, using the fibers as steps, you begin to climb upward. After climbing for some time and finding that you still haven't

_____  6.  reached the upper regions of the skin, you stop for a rest. As you sit, a strange-looking cell approaches, moving slowly

_____  7.  with parts alternately flowing forward and then receding. Suddenly you realize that this must be a __(4)__ that is about

_____  8.  to dispose of an intruder (you) unless you move in a hurry! You scramble to your feet and resume your upward climb.

_____  9.  On your right is a large fibrous structure that looks like a tree trunk anchored in place by muscle fibers. By scurrying up

_____  10.  this __(5)__ sheath, you are able to escape from the cell. Once safely out of harm's way, you again scan your surroundings.

Directly overhead are tall cubelike cells, forming a continuous sheet. In your rush to escape you have reached the __(6)__ layer of the skin. As you watch the activity of the cells in this layer, you notice that many of the cells are pinching in two, and that the daughter cells are being forced upward. Obviously, this is the layer that continually replaces cells that rub off the skin surface, and these cells are the __(7)__ cells.

Looking through the transparent cell membrane of one of the basal cells, you see a dark mass hanging over its nucleus. You wonder if this cell could have a tumor; but then, looking through the membranes of the neighboring cells, you find that they also have dark umbrella-like masses hanging over their nuclei. As you consider this matter, a black cell with long tentacles begins to pick its way carefully between the other cells. As you watch with interest, one of the transparent cells engulfs the tip of one of the black cell's tentacles. Within seconds a black substance appears above the transparent cell's nucleus. Suddenly, you remember that one of the skin's functions is to protect the deeper layers from sun damage; the black substance must be the protective pigment, __(8)__.

Once again you begin your upward climb and notice that the cells are becoming shorter and harder and are full of a waxy-looking substance. This substance has to be __(9)__, which would account for the increasing hardness of the cells. Climbing still higher, the cells become flattened like huge shingles. The only material apparent in the cells is the waxy substance—there is no nucleus, and there appears to be no activity in these cells. Considering the clues—shinglelike cells, no nuclei, full of the waxy substance, no activity—these cells are obviously __(10)__ and therefore very close to the skin surface.

Suddenly, you feel a strong agitation in your immediate area. The pressure is tremendous. Looking upward through the transparent cell layers, you see your host's fingertips vigorously scratching the area directly overhead. You wonder if you are causing his skin to sting or tickle. Then, within seconds, the cells around you begin to separate and fall apart, and you are catapulted out into the sunlight. Since the scratching fingers might descend once again, you quickly advise your host of your whereabouts.

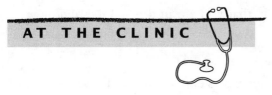

**18.** Mrs. Ibañez volunteered to help at a hospital for children with cancer. When she first entered the cancer ward, she was upset by the fact that most of the children had no hair. What is the explanation for their baldness?

**19.** A new mother brings her infant to the clinic, worried about a yellowish, scummy deposit that has built up on the baby's scalp. What is this condition called, and is it serious?

**20.** Patients in hospital beds are rotated every 2 hours to prevent bedsores. Exactly why is this effective?

**21.** Count Dracula, the most famous vampire of legends, was based on a real person who lived in eastern Europe about 600 years ago. He killed at least 200,000 people in the region he ruled. He was indeed a "monster," even though he was not a real vampire. Let your fingers do the walking through a medical dictionary and determine which of the following conditions the historical Count Dracula may have suffered from: (a) porphyria, (b) EB, (c) halitosis, (d) dermatitis.

**22.** After studying the skin in anatomy class, Toby grabbed the large "love handles" at his waist and said, "I have too thick a hypodermis, but that's OK because this layer performs some valuable functions!" What are the functions of the hypodermis?

**23.** A man got his finger caught in a machine at the factory. The damage was less serious than expected, but nonetheless, the entire nail was torn from his right index finger. The parts lost were the body, root, bed, matrix, and cuticle of the nail. First, define each of these parts. Then, tell if this nail is likely to grow back.

**24.** In cases of a ruptured appendix, what serous membrane is likely to become infected? Why can this be life-threatening?

# 5

# The Skeletal System

The skeleton is constructed of two of the most supportive tissues found in the human body—cartilage and bone. Besides supporting and protecting the body as an internal framework, the skeleton provides a system of levers that the skeletal muscles use to move the body. In addition, the bones provide a storage depot for substances such as lipids and calcium, and blood cell formation goes on within their red marrow cavities.

The skeleton consists of bones connected at joints, or articulations, and is subdivided into two divisions. The axial skeleton includes those bones that lie around the body's center of gravity. The appendicular skeleton includes the bones of the limbs.

Topics for student review include structure and function of long bones, location and naming of specific bones in the skeleton, fracture types, and a classification of joint types in the body.

## BONES—AN OVERVIEW

1. Classify each of the following terms as a projection (*P*) or a depression or opening (*D*). Enter the appropriate letter in the answer blanks.

    _____ 1. Condyle         _____ 4. Foramen         _____ 7. Ramus

    _____ 2. Crest           _____ 5. Head            _____ 8. Spine

    _____ 3. Fissure         _____ 6. Meatus          _____ 9. Tuberosity

2. Group each of the following bones into one of the four major bone categories. Use *L* for long bone, *S* for short bone, *F* for flat bone, and *I* for irregular bone. Enter the appropriate letter in the space provided.

    _____ 1. Calcaneus       _____ 4. Humerus         _____ 7. Radius

    _____ 2. Frontal         _____ 5. Mandible        _____ 8. Sternum

    _____ 3. Femur           _____ 6. Metacarpal      _____ 9. Vertebra

**3.** Using the key choices, characterize the following statements relating to long bones. Enter the appropriate term(s) or letter(s) in the answer blanks.

*Key Choices*

A. Diaphysis          C. Epiphysis          E. Yellow marrow cavity

B. Epiphyseal plate   D. Red marrow

_____ 1. Site of spongy bone in the adult

_____ 2. Site of compact bone in the adult

_____ 3. Site of hematopoiesis in the adult

_____ 4. Scientific name for bone shaft

_____ 5. Site of fat storage in the adult

_____ 6. Site of longitudinal growth in a child

**4.** Complete the following statements concerning bone formation and destruction, using the terms provided in the key. Insert the key letter or corresponding term in the answer blanks.

*Key Choices*

A. Atrophy       C. Gravity        E. Osteoclasts     G. Parathyroid hormone

B. Calcitonin    D. Osteoblasts    F. Osteocytes      H. Stress and/or tension

_____ 1. When blood calcium levels begin to drop below homeostatic levels, __(1)__ is released, causing calcium to be released from bones.

_____ 2. Mature bone cells, called __(2)__, maintain bone in a viable state.

_____ 3. Disuse such as that caused by paralysis or severe lack of exercise results in muscle and bone __(3)__.

_____ 4. Large tubercles and/or increased deposit of bony matrix occur at sites of __(4)__.

_____ 5. Immature, or matrix-depositing, bone cells are referred to as __(5)__.

_____ 6. __(6)__ causes blood calcium to be deposited in bones as calcium salts.

_____ 7. Bone cells that liquefy bone matrix and release calcium to the blood are called __(7)__.

_____ 8. Our astronauts must do isometric exercises when in space because bones atrophy under conditions of weightlessness or lack of __(8)__.

**5.** Five descriptions of bone structure are provided in Column A. First identify the structure by choosing the appropriate term from Column B and placing the corresponding answer in the answer blank. Then consider Figure 5–1A, a diagrammatic view of a cross section of bone, and 5–1B, a higher magnificated view of compact bone tissue. Select different colors for the structures and bone areas in Column B, and use them to color the coding circles and corresponding structures on the figure diagrams. Since the concentric lamellae would be difficult to color without confusing other elements, identify one lamella by using a bracket and label.

**Column A**

_____ 1. Layers of calcified matrix

_____ 2. "Residences" of osteocytes

_____ 3. Longitudinal canal, carrying blood vessels and nerves

_____ 4. Nonliving, structural part of bone

_____ 5. Tiny canals, connecting lacunae

**Column B**

A. Central (Haversian) canal ◯

B. Concentric lamellae

C. Lacunae ◯

D. Canaliculi ◯

E. Bone matrix ◯

F. Osteocyte ◯

A

B

**Figure 5–1**

**6.** Circle the term that does not belong in each of the following groupings.

1. Hematopoiesis       Red marrow       Yellow marrow       Spongy bone

2. Lamellae       Canaliculi       Circulation       Osteoblasts

3. Osteon       Marrow cavity       Central canal       Canaliculi

4. Epiphysis surface       Articular cartilage       Periosteum       Hyaline cartilage

7. Figure 5–2A is a midlevel, cross-sectional view of the diaphysis of the femur. Label the membrane that lines the cavity and the membrane that covers the outside surface.

Figure 5–2B is a drawing of a longitudinal section of the femur. Color the bone tissue gold. Do *not* color the articular cartilage; leave it white. Select different colors for the bone regions listed at the coding circles below. Color the coding circles and the corresponding regions on the drawing. Complete Figure 5–2B by labeling compact bone and spongy bone.

◯ Diaphysis  ◯ Area where red marrow is found

◯ Epiphyseal plate  ◯ Area where yellow marrow is found

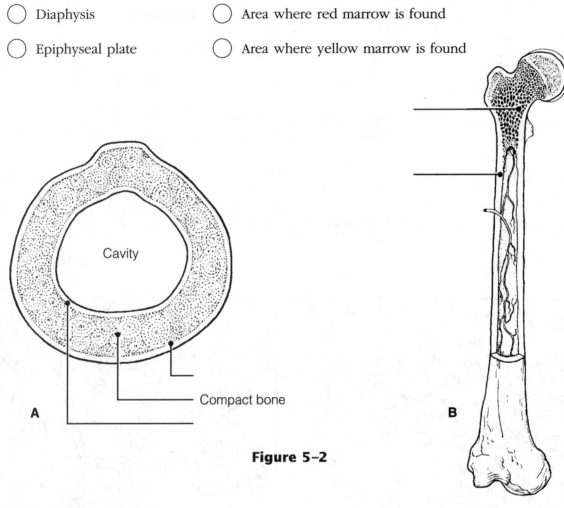

Cavity

Compact bone

A  B

**Figure 5–2**

# AXIAL SKELETON

## Skull

8. Using key choices, identify the bones indicated by the following descriptions. Enter the appropriate term or letter in the answer blanks.

*Key Choices*

A. Ethmoid  E. Mandible  I. Palatines  L. Temporals

B. Frontal  F. Maxillae  J. Parietals  M. Vomer

C. Hyoid  G. Nasals  K. Sphenoid  N. Zygomatic

D. Lacrimals  H. Occipital

_____ 1. Forehead bone

_____ 2. Cheekbone

_____ 3. Lower jaw

_____ 4. Bridge of nose

_____ 5. Posterior part of hard palate

_____ 6. Much of the lateral and superior cranium

_____ 7. Most posterior part of cranium

_____ 8. Single, irregular, bat-shaped bone, forming part of the cranial floor

_____ 9. Tiny bones, bearing tear ducts

_____ 10. Anterior part of hard palate

_____ 11. Superior and middle nasal conchae formed from its projections

_____ 12. Site of mastoid process

_____ 13. Site of sella turcica

_____ 14. Site of cribriform plate

_____ 15. Site of mental foramen

_____ 16. Site of styloid process

_____ 17. _____ 18. Four bones, containing paranasal sinuses

_____ 19. _____ 20.

_____ 21. Its condyles articulate with the atlas

_____ 22. Foramen magnum contained here

_____ 23. Middle ear found here

_____ 24. Nasal septum

_____ 25. Bears an upward protrusion, the "cock's comb," or crista galli

9. Figure 5–3, A–C shows lateral, anterior, and inferior views of the skull. Select different colors for the bones listed below and color the coding circles and corresponding bones in the figure. Complete the figure by labeling the bone markings indicated by leader lines.

◯ Frontal          ◯ Sphenoid          ◯ Zygomatic          ◯ Nasal

◯ Parietal          ◯ Ethmoid          ◯ Palatine          ◯ Lacrimal

◯ Mandible          ◯ Temporal          ◯ Occipital          ◯ Vomer

◯ Maxilla

A

**Figure 5–3, A–C**

B

C

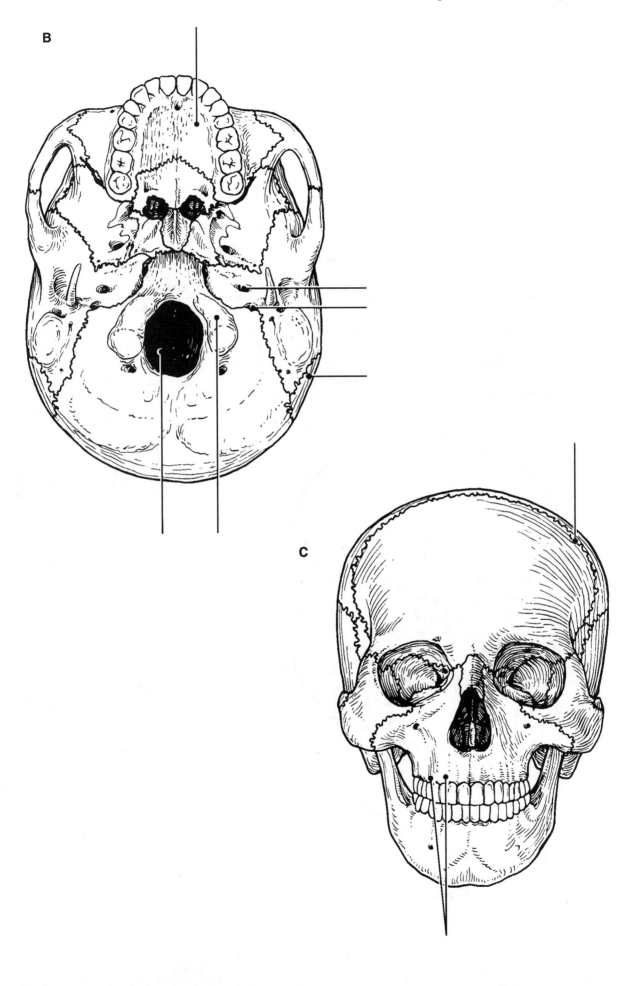

**10.** An anterior view of the skull, showing the positions of the sinuses, is provided in Figure 5–4. First select different colors for each of the sinuses and
use them to color the coding circles and the corresponding structures on the
figure. Then briefly answer the following questions concerning the sinuses.

1. What *are* sinuses? _____

2. What purpose do they serve in the skull? _____

_____

_____

3. Why are they so susceptible to infection? _____

_____

_____

◯ Sphenoid sinus        ◯ Ethmoid sinuses

◯ Frontal sinus         ◯ Maxillary sinus

**Figure 5–4**

# Vertebral Column

**11.** Using the key choices, correctly identify the vertebral parts/areas described as
follows. Enter the appropriate term(s) or letter(s) in the spaces provided.

*Key Choices*

A. Body                    C. Spinous process           E. Transverse process

B. Intervertebral foramina     D. Superior articular process     F. Vertebral arch

_____ 1. Structure that encloses the nerve cord

_____ 2. Weight-bearing portion of the vertebra

_____ 3. Provide(s) levers for the muscles to pull against

_____ 4. Provide(s) an articulation point for the ribs

_____ 5. Openings providing for exit of spinal nerves

**12.** The following statements provide distinguishing characteristics of the vertebrae composing the vertebral column. Using key choices, identify each
described structure or region by inserting the appropriate term(s) or letter(s)
in the spaces provided.

*Key Choices*

A. Atlas                    D. Coccyx              F. Sacrum

B. Axis                     E. Lumbar vertebra     G. Thoracic vertebra

C. Cervical vertebra—typical

_____ 1. Type of vertebra(e) containing foramina in the transverse processes,
through which the vertebral arteries ascend to reach the brain

_____ 2. Its dens provides a pivot for rotation of the first cervical vertebra

_____ 3. Transverse processes have facets for articulation with ribs;
spinous process points sharply downward

_____ 4. Composite bone; articulates with the hip bone laterally

_____ 5. Massive vertebrae; weight-sustaining

_____ 6. Tailbone; vestigal fused vertebrae

_____ 7. Supports the head; allows the rocking motion of the occipital
condyles

_____ 8. Seven components; unfused

_____ 9. Twelve components; unfused

**13.** Complete the following statements by inserting your answers in the answer blanks.

_____ 1.

_____ 2.

_____ 3.

_____ 4.

1. In describing abnormal curvatures, it could be said that __(1)__ is an exaggerated thoracic curvature, and in __(2)__ the verte-
2. bral column is displaced laterally.

3. Invertebral discs are made of __(3)__ tissue. The discs provide __(4)__ to the spinal column.

**14.** Figure 5–5, A–D shows superior views of four types of vertebrae. In the spaces provided below each vertebra, indicate in which region of the spinal column it would be found. In addition, specifically identify Figure 5–5A. Where indicated by leader lines, identify the vertebral body, spinous and transverse processes, superior articular processes, and vertebral foramen.

A _____        B _____

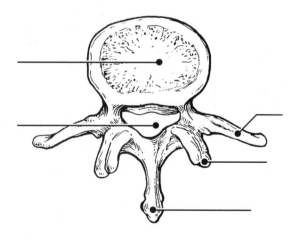

C _____        D _____

**Figure 5–5**

**15.** Figure 5–6 is a lateral view of the vertebral column. Identify each numbered region of the column by listing in the numbered answer blanks the region name first and then the specific vertebrae involved (for example, sacral region, S# to S#). Also identify the modified vertebrae indicated by numbers 6 and 7 in Figure 5–6. Select different colors for each vertebral region and use them to color the coding circles and the corresponding regions.

1. _____  ◯

2. _____  ◯

3. _____  ◯

4. _____  ◯

5. _____  ◯

6. _____  ◯

7. _____  ◯

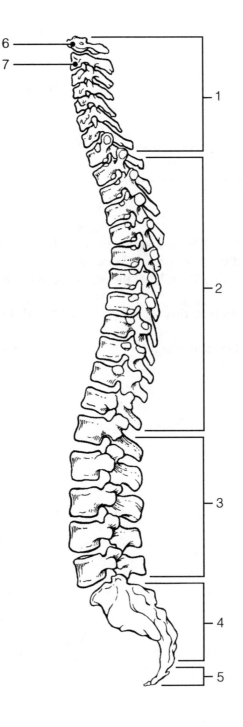

**Figure 5–6**

# Bony Thorax

**16.** Complete the following statements referring to the bony thorax by inserting
your responses in the answer blanks.

_____ 1.

_____ 2.

_____ 3.

_____ 4.

_____ 5.

_____ 6.

_____ 7.

_____ 8.

The organs protected by the thoracic cage include the __(1)__
and the __(2)__. Ribs 1 through 7 are called __(3)__ ribs,
whereas ribs 8 through 12 are called __(4)__ ribs. Ribs 11 and
12 are also called __(5)__ ribs. All ribs articulate posteriorly
with the __(6)__, and most connect anteriorly to the __(7)__,
either directly or indirectly.

The general shape of the thoracic cage is __(8)__.

**17.** Figure 5–7 is an anterior view of the bony thorax. Select different colors to
identify the structures below and color the coding circles and corresponding
structures. Then label the subdivisions of the sternum indicated by leader lines.

◯ All true ribs           ◯ All false ribs

◯ Costal cartilages       ◯ Sternum

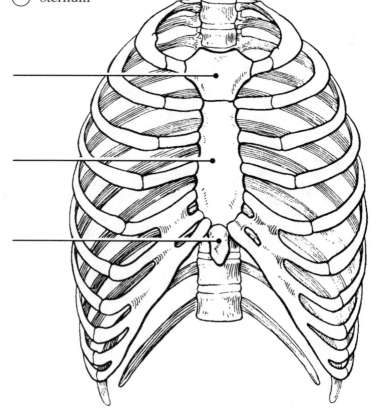

**Figure 5–7**

# APPENDICULAR SKELETON

Several bones forming part of the upper limb and/or shoulder girdle are shown in Figures 5–8 to 5–11. Follow the specific directions for each figure.

**18.** Identify the bone in Figure 5–8. Insert your answer in the blank below the illustration. Select different colors for each structure listed below and use them to color the coding circles and the corresponding structures in the diagram. Then, label the angles indicated by leader lines.

◯ Spine     ◯ Glenoid cavity     ◯ Coracoid process     ◯ Acromion

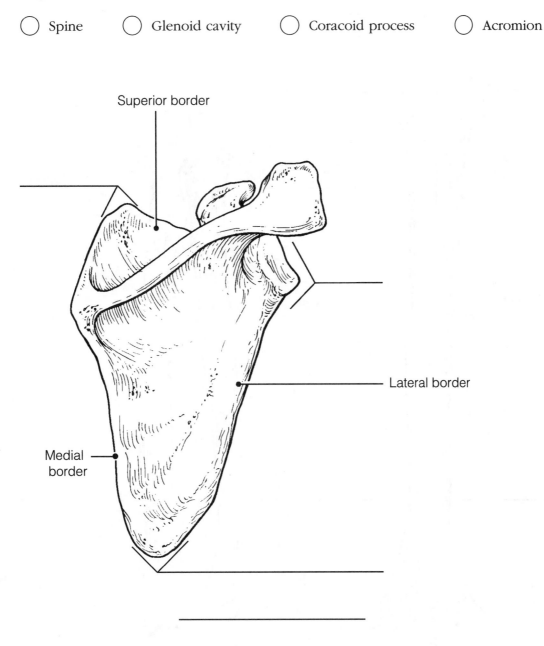

Superior border

Lateral border

Medial border

**Figure 5–8**

**19.** Identify the bones in Figure 5–9 by labeling the leader lines identified as A, B, and C. Color the bones different colors. Using the following terms, complete the illustration by labeling all bone markings provided with leader lines.

| | | |
|---|---|---|
| Trochlear notch | Capitulum | Coronoid process |
| Trochlea | Deltoid tuberosity | Olecranon process |
| Radial tuberosity | Head (three) | Greater tubercle |
| | Styloid process | Lesser tubercle |

**Figure 5–9**

**20.** Figure 5–10 is a diagram of the hand. Select different colors for the following structures, and use them to color the coding circles and the corresponding structures in the diagram.

◯ Carpals          ◯ Metacarpals          ◯ Phalanges

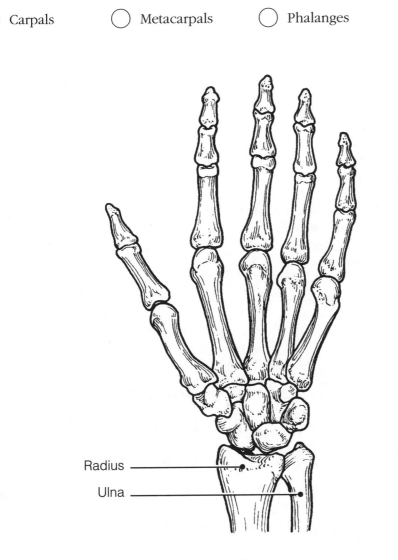

Radius —

Ulna —

**Figure 5-10**

**21.** Compare the pectoral and pelvic girdles by choosing descriptive terms from the key choices. Insert the appropriate key letters in the answer blanks.

***Key Choices***

A. Flexibility          D. Shallow socket for limb attachment

B. Massive             E. Deep, secure socket for limb attachment

C. Lightweight         F. Weight-bearing

Pectoral: _____, _____, _____          Pelvic: _____, _____, _____

**22.** Using key choices, identify the bone names or markings according to the descriptions that follow. Insert the appropriate term or letter in the answer blanks.

*Key Choices*

A. Acromion      F. Coronoid fossa      K. Olecranon fossa      P. Scapula

B. Capitulum      G. Deltoid tuberosity      L. Olecranon process      Q. Sternum

C. Carpals      H. Glenoid cavity      M. Phalanges      R. Styloid process

D. Clavicle      I. Humerus      N. Radial tuberosity      S. Trochlea

E. Coracoid process      J. Metacarpals      O. Radius      T. Ulna

_____ 1. Raised area on lateral surface of humerus to which deltoid muscle attaches

_____ 2. Arm bone

_____ 3. _____ 4. Bones composing the shoulder girdle

_____ 5. _____ 6. Forearm bones

_____ 7. Point where scapula and clavicle connect

_____ 8. Shoulder girdle bone that has no attachment to the axial skeleton

_____ 9. Shoulder girdle bone that articulates anteriorly with the sternum

_____ 10. Socket in the scapula for the arm bone

_____ 11. Process above the glenoid cavity that permits muscle attachment

_____ 12. Commonly called the collarbone

_____ 13. Distal medial process of the humerus; joins the ulna

_____ 14. Medial bone of the forearm in anatomical position

_____ 15. Rounded knob on the humerus that articulates with the radius

_____ 16. Anterior depression; superior to the trochlea; receives part of the ulna when the forearm is flexed

_____ 17. Forearm bone involved in formation of elbow joint

_____ 18. _____ 19. Bones that articulate with the clavicle

_____ 20. Bones of the wrist

_____ 21. Bones of the fingers

_____ 22. Heads of these bones form the knuckles

**23.** Figure 5–11 is a diagram of the articulated pelvis. Identify the bones and bone markings indicated by leader lines on the figure. Select different colors for the structures listed below and use them to color the coding circles and the corresponding structures in the figure. Also, label the dashed line showing the dimensions of the true pelvis and that showing the diameter of the false pelvis. Complete the illustration by labeling the following bone markings: obturator foramen, iliac crest, anterior superior iliac spine, ischial spine, pubic ramus, and pelvic brim. Last, list three ways in which the female pelvis differs from the male pelvis and insert your answers in the answer blanks.

◯ Coxal bone    ◯ Pubic symphysis

◯ Sacrum    ◯ Acetabulum

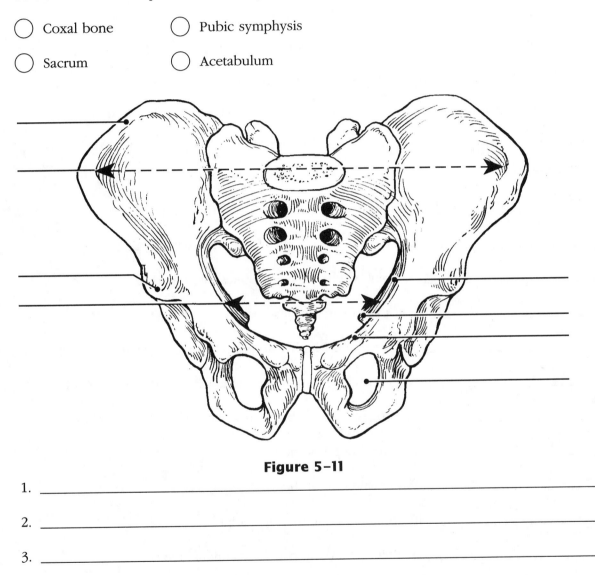

**Figure 5–11**

1. _____

2. _____

3. _____

**24.** Circle the term that does not belong in each of the following groupings.

1. Tibia    Ulna    Fibula    Femur

2. Skull    Rib cage    Vertebral column    Pelvis

3. Ischium    Scapula    Ilium    Pubis

4. Mandible    Frontal bone    Temporal bone    Occipital bone

5. Calcaneus    Tarsals    Carpals    Talus

**25.** Using key choices, identify the bone names and markings, according to the descriptions that follow. Insert the appropriate key term(s) or letter(s) in the answer blanks.

***Key Choices***

A. Acetabulum

B. Calcaneus

C. Femur

D. Fibula

E. Gluteal tuberosity

F. Greater sciatic notch

G. Greater and lesser trochanters

H. Iliac crest

I. Ilium

J. Ischial tuberosity

K. Ischium

L. Lateral malleolus

M. Lesser sciatic notch

N. Medial malleolus

O. Metatarsals

P. Obturator foramen

Q. Patella

R. Pubic symphysis

S. Pubis

T. Sacroiliac joint

U. Talus

V. Tarsals

W. Tibia

X. Tibial tuberosity

_____ 1. Fuse to form the coxal bone (hip bone)

_____ 2. Receives the weight of the body when sitting

_____ 3. Point where the coxal bones join anteriorly

_____ 4. Upper margin of iliac bones

_____ 5. Deep socket in the hip bone that receives the head of the thigh bone

_____ 6. Point where axial skeleton attaches to the pelvic girdle

_____ 7. Longest bone in body, articulates with the coxal bone

_____ 8. Lateral bone of the leg

_____ 9. Medial bone of the leg

_____ 10. Bones forming the knee joint

_____ 11. Point where the patellar ligament attaches

_____ 12. Kneecap

_____ 13. Shinbone

_____ 14. Distal process on medial tibial surface

_____ 15. Process forming the outer ankle

_____ 16. Heel bone

_____ 17. Bones of the ankle

_____ 18. Bones forming the instep of the foot

_____ 19. Opening in a coxal bone formed by the pubic and ischial rami

_____ 20. Sites of muscle attachment on the proximal end of the femur

_____ 21. Tarsal bone that articulates with the tibia

**26.** For each of the following statements that is true, insert *T* in the answer blank. If any of the statements are false, correct the underlined term by inserting the correct term in the answer blank.

_____ 1. The <u>pectoral</u> girdle is formed by the articulation of the hip bones and the sacrum.

_____ 2. Bones present in both the hand and the foot are <u>carpals</u>.

_____ 3. The tough, fibrous connective tissue covering of a bone is the <u>periosteum</u>.

_____ 4. The point of fusion of the three bones forming a coxal bone is the <u>glenoid cavity</u>.

_____ 5. The large nerve that must be avoided when giving injections into the buttock muscles is the <u>femoral</u> nerve.

_____ 6. The long bones of a fetus are constructed of <u>hyaline</u> cartilage.

_____ 7. Bones that provide the most protection to the abdominal viscera are the <u>ribs</u>.

_____ 8. The largest foramen in the skull is the <u>foramen magnum</u>.

**27.** The bones of the thigh and the leg are shown in Figure 5–12. Identify each and put your answers in the blanks labelled A, B, and C. Select different colors for the lower limb bones listed below and use them to color in the coding circles and corresponding bones on the diagram. Complete the illustration by inserting the terms indicating bone markings at the ends of the appropriate leader lines in the figure.

◯ Femur                    ◯ Tibia                    ◯ Fibula

Head of femur              Anterior crest of tibia          Head of fibula

Intercondylar eminence     Lesser trochanter               Medial malleolus

Tibial tuberosity          Greater trochanter              Lateral malleolus

**Figure 5–12**

**28.** Figure 5–13 is a diagram of the articulated skeleton. Identify all bones or groups of bones by writing the correct labels at the end of the leader lines. Then, select two different colors for the bones of the axial and appendicular skeletons and use them to color in the coding circles and corresponding structures in the diagram.

◯ Axial skeleton          ◯ Appendicular skeleton

**Figure 5–13**

# BONE FRACTURES

**29.** Using the key choices, identify the fracture (fx) types shown in Figure 5–14 and the fracture types and treatments described below. Enter the appropriate key letter or term in each answer blank.

### *Key Choices*

A. Closed reduction      D. Depressed fracture      G. Simple fracture

B. Compression fracture      E. Greenstick fracture      H. Spiral fracture

C. Compound fracture      F. Open reduction

_____ 1. Bone is broken cleanly; the ends do not penetrate the skin

_____ 2. Nonsurgical realignment of broken bone ends and splinting of bone

_____ 3. A break common in children; bone splinters, but break is incomplete

_____ 4. A fracture in which the bone is crushed; common in the vertebral column

_____ 5. A fracture in which the bone ends penetrate through the skin surface

_____ 6. Surgical realignment of broken bone ends

_____ 7. A result of twisting forces

**Figure 5–14**

**30.** For each of the following statements about bone breakage and the repair process that is true, insert *T* in the answer blank. For false statements, correct the <u>underlined</u> terms by inserting the correct term in the answer blank.

_____ 1. A <u>hematoma</u> usually forms at a fracture site.

_____ 2. Deprived of nutrition, <u>osteocytes</u> at the fracture site die.

_____ 3. Non-bony debris at the fracture site is removed by <u>osteoclasts</u>.

_____ 4. Growth of a new capillary supply into the region produces <u>granulation tissue</u>.

_____ 5. Osteoblasts from the <u>medullary cavity</u> migrate to the fracture site.

_____ 6. The <u>fibrocartilage callus</u> is the first repair mass to splint the broken bone.

_____ 7. The bony callus is initially composed of <u>compact</u> bone.

# JOINTS

**31.** Figure 5–15 shows the structure of a typical diarthrotic joint. Select different colors to identify each of the following areas and use them to color the coding circles and the corresponding structures on the figure. Then, complete the statements below the figure.

○ Articular cartilage of bone ends

○ Fibrous capsule

○ Synovial membrane

○ Joint cavity

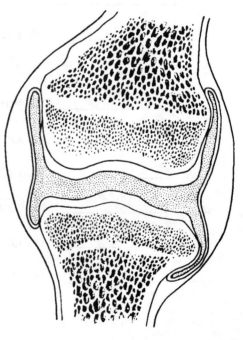

**Figure 5–15**

1. _____ The lubricant that minimizes friction and abrasion of joint surfaces is __(1)__ .

2. _____ The resilient substance that keeps bone ends from crushing when compressed is __(2)__ .

3. _____ __(3)__ which reinforce the fibrous capsule help to prevent dislocation of the joint.

**32.** For each joint described below, select an answer from Key A. Then, if the Key A selection *is other than C* (a synovial joint), see if you can classify the joint further by making a choice from Key B.

**Key Choices**

Key A:    A. Cartilaginous          Key B:    1. Epiphyseal disk

            B. Fibrous                         2. Suture

            C. Synovial                         3. Symphysis

_____    1. Has amphiarthrotic and synarthrotic examples

_____    2. All have a fibrous capsule lined with synovial membrane surrounding a joint cavity

_____    3. Bone regions united by fibrous connective tissue

_____    4. Joints between skull bones

_____    5. Joint between atlas and axis

_____    6. Hip, elbow, and knee

_____    7. All examples are diarthroses

_____    8. Pubic symphysis

_____    9. All are reinforced by ligaments

_____    10. Joint providing the most protection to underlying structures

_____    11. Often contains a fluid-filled cushion

_____    12. Child's long-bone growth plate made of hyaline cartilage

_____    13. Most joints of the limbs

_____    14. Often associated with bursae

_____    15. Have the greatest mobility

**33.** Which structural joint type is *not* commonly found in the axial skeleton and why not?

_____

_____

_____

## Homeostatic Imbalances of Bones and Joints

**34.** For each of the following statements that is true, enter *T* in the answer blank. For each false statement, correct the <u>underlined</u> words by writing the correct words in the answer blank.

_____ 1. In a <u>sprain</u>, the ligaments reinforcing a joint are excessively stretched or torn.

_____ 2. Age-related erosion of articular cartilages and formation of painful bony spurs are characteristic of <u>gouty arthritis</u>.

_____ 3. <u>Chronic</u> arthritis usually results from bacterial invasion.

_____ 4. Healing of a partially torn ligament is slow because its hundreds of fibrous strands are poorly <u>aligned</u>.

_____ 5. <u>Rheumatoid arthritis</u> is an autoimmune disease.

_____ 6. High levels of uric acid in the blood may lead to <u>rheumatoid arthritis</u>.

_____ 7. A "soft" bone condition in children, usually due to a lack of calcium or vitamin D in the diet, is called <u>osteomyelitis</u>.

_____ 8. Atrophy and thinning of bone owing to hormonal changes or inactivity (generally in the elderly) is called <u>osteoporosis</u>.

# DEVELOPMENTAL ASPECTS OF THE SKELETON

**35.** Using the key choices, identify the body systems that relate to bone tissue viability. Enter the appropriate key terms or letters in the answer blanks.

### Key Choices

A. Endocrine          C. Muscular          E. Reproductive

B. Integumentary      D. Nervous           F. Urinary

_____ 1. Conveys the sense of pain in bone and joints

_____ 2. Activates vitamin D for proper calcium usage

_____ 3. Regulates uptake and release of calcium by bones

_____ 4. Increases bone strength and viability by pulling action

_____ 5. Influences skeleton proportions and adolescent growth of long bones

_____ 6. Provides vitamin D for proper calcium absorption

**36.** Complete the following statements concerning fetal and infant skeletal development. Insert the missing words in the answer blanks.

_____ 1.

_____ 2.

_____ 3.

_____ 4.

_____ 5.

_____ 6.

_____ 7.

_____ 8.

_____ 9.

"Soft spots," or membranous joints called __(1)__ in the fetal skull, allow the skull to be __(2)__ slightly during birth passage. They also allow for continued brain __(3)__ during the later months of fetal development and early infancy. Eventually these soft spots are replaced by immovable joints called __(4)__.

The two spinal curvatures well developed at birth are the __(5)__ and __(6)__ curvatures. Because they are present at birth, they are called __(7)__ curvatures. The secondary curvatures develop as the baby matures. The __(8)__ curvature develops as the baby begins to lift his or her head. The __(9)__ curvature matures when the baby begins to walk or assume the upright posture.

# INCREDIBLE JOURNEY

## A Visualization Exercise for the Skeletal System

_... stalagmite- and stalactite-like structures that surround you. ..._
_Since the texture is so full of holes..._

**37.** Where necessary, complete statements by inserting the missing words in the answer blanks.

_____ 1.

_____ 2.

_____ 3.

_____ 4.

_____ 5.

_____ 6.

For this journey you are miniaturized and injected into the interior of the largest bone of your host's body, the __(1)__. Once inside this bone, you look around and find yourself examining the stalagmite- and stalactite-like structures that surround you. Although you feel as if you are in an underground cavern, you know that it has to be bone. Since the texture is so full of holes, it obviously is __(2)__ bone. Although the arrangement of these bony spars seems to be haphazard, as if someone randomly dropped straws, they are precisely arranged to resist points of __(3)__. All about you is frantic, hurried activity. Cells are dividing rapidly, nuclei are being ejected, and disk-like cells are appearing. You decide that these disk-like cells are __(4)__, and that this is the __(5)__ cavity. As you explore further, strolling along the edge of the cavity, you spot many tunnels leading into the solid bony area on which you are walking. Walking into one of these drainpipe-like openings, you notice that it contains a glistening white rope-like structure (a __(6)__, no doubt), and blood vessels running the length of the tube. You eventually come to a point in the channel where

_____ 7.

_____ 8.

_____ 9.

_____ 10.

_____ 11.

_____ 12.

the horizontal passageway joins with a vertical passage that runs with the longitudinal axis of the bone. This is obviously a __(7)__ canal. Since you would like to see how nutrients are brought into __(8)__ bone, you decide to follow this channel. Reasoning that there is no way you can possibly scale the slick walls of the channel, you leap and grab onto a white cord hanging down its length. Since it is easier to slide down than to try to climb up the cord, you begin to lower yourself, hand over hand. During your descent, you notice small openings in the wall, which are barely large enough for you to wriggle through. You conclude that these are the __(9)__ that connect all the __(10)__ to the nutrient supply in the central canal. You decide to investigate one of these tiny openings and begin to swing on your cord, trying to get a foothold on one of the openings. After managing to anchor yourself and squeezing into an opening, you use a flashlight to illuminate the passageway in front of you. You are startled by a giant cell with many dark nuclei. It appears to be plastered around the entire lumen directly ahead of you. As you watch this cell, the bony material beneath it, the __(11)__ , begins to liquefy. The cell apparently is a bone-digesting cell, or __(12)__ , and since you are unsure whether or not its enzymes can also liquefy you, you slither backwards hurriedly and begin your trek back to your retrieval site.

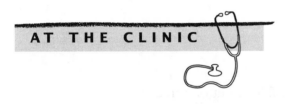

**AT THE CLINIC**

**38.** Antonio is hit in the face with a football during practice. An X-ray reveals multiple fractures of the bones around an orbit. Name the bones that form margins of the orbit.

**39.** Mrs. Bruso, a woman in her 80s, is brought to the clinic with a fractured hip. X-rays reveal compression fractures in her lower vertebral column and extremely low bone density in her vertebrae, hip bones, and femurs. What are the condition, cause, and treatment?

**40.** Jack, a young man, is treated at the clinic for an accident in which he hit his forehead. When he returns for a checkup, he complains that he can't smell anything. A hurried X-ray of his head reveals a fracture. What part of which bone was fractured to cause his loss of smell?

**41.** A middle-aged woman comes to the clinic complaining of stiff, painful joints and increasing immobility of her finger joints. A glance at her hands reveals knobby, deformed knuckles. What condition will be tested for?

**42.** At his 94th birthday party, James was complimented on how good he looked and was asked about his health. He replied, "I feel good most of the time, but some of my joints ache and are stiff, especially my knees, hips, and lower back, and especially in the morning when I wake up." A series of X-rays and an MRI scan taken a few weeks earlier had revealed that the articular cartilages of these joints was rough and flaking off, and that bone spurs (overgrowths) were present at the ends of some of James's bones. What is James's probable condition?

**43.** Janet, a 10-year-old girl, is brought to the clinic after falling out of a tree. An X-ray shows she has small fractures of the transverse processes of T3 to T5 on the right side. Janet will be watched for what abnormal spinal curvature over the next several years?

# The Muscular System

Muscles, the specialized tissues that facilitate body movement, make up about 40% of body weight. Most body muscle is the voluntary type, called skeletal muscle because it is attached to the bony skeleton. Skeletal muscle contributes to body contours and shape, and composes the organ system called the muscular system. These muscles allow you to grin, frown, run, swim, shake hands, swing a hammer, and to otherwise manipulate your environment. The balance of body muscle is smooth and cardiac muscles, which form the bulk of the walls of hollow organs and the heart. Smooth and cardiac muscles are involved in the transport of materials within the body.

Study activities in this chapter deal with microscopic and gross structure of muscle, identification of voluntary muscles, body movements, and important understandings of muscle physiology.

## OVERVIEW OF MUSCLE TISSUES

**1.** Nine characteristics of muscle tissue are listed below and on page 90. Identify the muscle tissue type described by choosing the correct response(s) from the key choices. Enter the appropriate term(s) or letter(s) of the key choice in the answer blank.

### Key Choices

A. Cardiac          B. Smooth          C. Skeletal

_____ 1. Involuntary

_____ 2. Banded appearance

_____ 3. Longitudinally and circularly arranged layers

_____ 4. Dense connective tissue packaging

_____ 5. Figure-8 packaging of the cells

_____ 6. Coordinated activity to act as a pump

*Key Choices*

A. Cardiac          B. Smooth          C. Skeletal

_____ 7. Moves bones and the facial skin

_____ 8. Referred to as the muscular system

_____ 9. Voluntary

2. Identify the type of muscle in each of the illustrations in Figure 6–1. Color the diagrams as you wish.

A ———————————————

Intercalated disks ⎯

B ———————————————

**Figure 6–1**

3. Regarding the functions of muscle tissues, circle the term in each of the groupings that does not belong with the other terms.

1.  Urine          Foodstuffs          Bones          Smooth muscle

2.  Heart          Cardiac muscle          Blood pump          Promotes labor during birth

3.  Excitability          Response to a stimulus          Contractility          Action potential

4.  Ability to shorten          Contractility          Pulls on bones          Stretchability

5.  Maintains posture          Movement          Promotes growth          Generates heat

# MICROSCOPIC ANATOMY OF SKELETAL MUSCLE

**4.** First, identify the structures in Column B by matching them with the descriptions in Column A. Enter the correct letters (or terms if desired) in the answer blanks. Then, select a different color for each of the terms in Column B that has a color-coding circle and color the structures in on Figure 6–2.

**Column A**

**Column B**

_____ 1. Connective tissue surrounding a fascicle

A. Endomysium ◯

_____ 2. Connective tissue ensheathing the entire muscle

B. Epimysium ◯

C. Fascicle

_____ 3. Contractile unit of muscle

D. Fiber ◯

_____ 4. A muscle cell

E. Myofilament

_____ 5. Thin connective tissue investing each muscle cell

F. Myofibril ◯

_____ 6. Plasma membrane of the muscle cell

G. Perimysium ◯

H. Sarcolemma

_____ 7. A long, filamentous organelle found within muscle cells that has a banded appearance

I. Sarcomere

J. Sarcoplasm

_____ 8. Actin- or myosin-containing structure

K. Tendon ◯

_____ 9. Cordlike extension of connective tissue beyond the muscle, serving to attach it to the bone

_____ 10. A discrete bundle of muscle cells

**Figure 6–2**

**5.** Figure 6–3 is a diagrammatic representation of a small portion of a relaxed muscle cell (bracket indicates the portion enlarged). First, select different colors for the structures listed below. Use them to color the coding circles and corresponding structures on Figure 6–3. Then bracket and label an A band, an I band, and a sarcomere. When you have finished, draw a contracted sarcomere in the space beneath the figure and label the same structures.

◯ Myosin          ◯ Actin filaments          ◯ Z disc

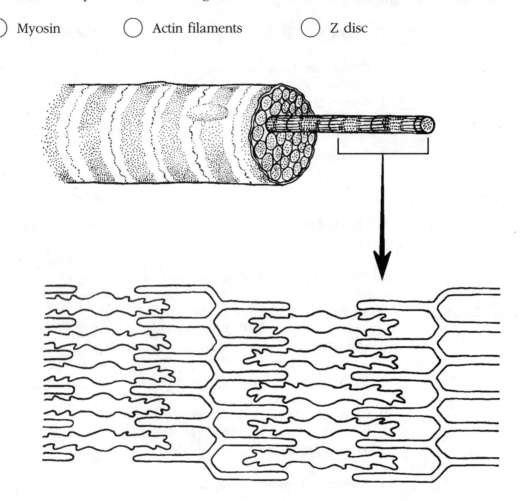

**Figure 6–3**

# SKELETAL MUSCLE ACTIVITY

**6.** Complete the following statements relating to the neuromuscular junction. Insert the correct answers in the numbered answer blanks.

_____ 1.

_____ 2.

_____ 3.

_____ 4.

_____ 5.

_____ 6.

A motor neuron and all of the skeletal muscle cells it stimulates is called a __(1)__. The axon of each motor neuron has numerous endings called __(2)__. The actual gap between an axonal ending and the muscle cell is called a __(3)__. Within the axonal endings are many small vesicles containing a neurotransmitter substance called __(4)__.

When the __(5)__ reaches the ends of the axon, the neurotransmitter is released, and it diffuses to the muscle cell membrane to combine with receptors there. Binding of the neurotransmitters with muscle membrane receptors causes the membrane to become permeable to sodium, resulting in the influx of sodium ions and __(6)__ of the membrane. Then contraction of the muscle cell occurs.

**7.** Figure 6–4 shows the components of a neuromuscular junction. Identify the parts by coloring the coding circles and the corresponding structures in the diagram. Add small arrows to indicate the location of the ACh receptors and label appropriately.

◯ Mitochondrion    ◯ T tubule    ◯ Sarcomere

◯ Synaptic vesicles    ◯ Synaptic cleft

**Figure 6–4**

8. Number the following statements in their proper sequence to describe the contraction mechanism in a skeletal muscle cell. The first step has already been identified as number 1.

    __1__    1. Acetylcholine is released into the neuromuscular junction by the axonal terminal.

    _____   2. The action potential, carried deep into the cell, causes the sarcoplasmic reticulum to release calcium ions.

    _____   3. The muscle cell relaxes and lengthens.

    _____   4. Acetylcholine diffuses across the neuromuscular junction and binds to receptors on the sarcolemma.

    _____   5. The calcium ion concentration at the myofilaments increases; the myofilaments slide past one another, and the cell shortens.

    _____   6. Depolarization occurs, and the action potential is generated.

    _____   7. As calcium is actively reabsorbed into the sarcoplasmic reticulum, its concentration at the myofilaments decreases.

9. The following incomplete statements refer to a muscle cell in the resting, or polarized, state just before stimulation. Complete each statement by choosing the correct response from the key choices and entering the appropriate letter in the answer blanks.

    **Key Choices**

    A. $Na^+$ diffuses out of the cell

    B. $K^+$ diffuses out of the cell

    C. $Na^+$ diffuses into the cell

    D. $K^+$ diffuses into the cell

    E. Inside the cell

    F. Outside the cell

    G. Relative ionic concentrations on the two sides of the membrane during rest

    H. Electrical conditions

    I. Activation of the sodium-potassium pump, which moves $K^+$ into the cell and $Na^+$ out of the cell

    J. Activation of the sodium-potassium pump, which moves $Na^+$ into the cell and $K^+$ out of the cell

    _____ 1.

    _____ 2.

    _____ 3.

    _____ 4.

    _____ 5.

    _____ 6.    _____ 7.

    There is a greater concentration of $Na^+$ __(1)__ , and there is a greater concentration of $K^+$ __(2)__ . When the stimulus is delivered, the permeability of the membrane is changed, and __(3)__ , initiating the depolarization of the membrane. Almost as soon as the depolarization wave begins, a repolarization wave follows it across the membrane. This occurs as __(4)__ . Repolarization restores the __(5)__ of the resting cell membrane. The __(6)__ is (are) reestablished by __(7)__ .

**10.** Complete the following statements by choosing the correct response from the key choices and entering the appropriate letter or term in the answer blanks.

*Key Choices*

A. Fatigue          E. Isometric contraction      I. Many motor units

B. Isotonic contraction    F. Whole muscle          J. Repolarization

C. Muscle cell        G. Tetanus            K. Depolarization

D. Muscle tone        H. Few motor units

_____ 1. _____ is a continuous contraction that shows no evidence of relaxation.

_____ 2. A(n) _____ is a contraction in which the muscle shortens and work is done.

_____ 3. To accomplish a strong contraction, _____ are stimulated at a rapid rate.

_____ 4. When a weak but smooth muscle contraction is desired, _____ are stimulated at a rapid rate.

_____ 5. When a muscle is being stimulated but is not able to respond due to "oxygen debt," the condition is called _____.

_____ 6. A(n) _____ is a contraction in which the muscle does not shorten but tension in the muscle keeps increasing.

**11.** The terms in the key refer to the three ways that muscle cells replenish their ATP supplies. Select the term(s) that best apply to the conditions described and insert the correct key letter(s) in the answer blanks.

*Key Choices*

A. Coupled reaction of CP and ADP        C. Aerobic respiration

B. Anaerobic glycolysis

_____ 1. Accompanied by lactic acid formation

_____ 2. Supplies the highest ATP yield per glucose molecule

_____ 3. Involves the simple transfer of a phosphate group

_____ 4. Requires no oxygen

_____ 5. The slowest ATP regeneration process

_____ 6. Produces carbon dioxide and water

_____ 7. The energy mechanism used in the second hour of running in a marathon

_____ 8. Used when the oxygen supply is inadequate over time

_____ 9. Good for a sprint

**12.** Briefly describe how you can tell when you are repaying the oxygen debt.

_____

_____

**13.** Which of the following occur within a muscle cell during oxygen debt? Place a check (✓) by the correct choices.

_____    1. Decreased ATP            _____    5. Increased oxygen

_____    2. Increased ATP            _____    6. Decreased carbon dioxide

_____    3. Increased lactic acid    _____    7. Increased carbon dioxide

_____    4. Decreased oxygen         _____    8. Increased glucose

# MUSCLE MOVEMENTS, TYPES, AND NAMES

**14.** Relative to general terminology concerning muscle activity, first label the following structures on Figure 6–5: insertion, origin, tendon, resting muscle, and contracting muscle. Next, identify the two structures named below by choosing different colors for the coding circles and the corresponding structures in the figure.

◯ Movable bone

◯ Immovable bone

**Figure 6–5**

**15.** Complete the following statements. Insert your answers in the answer blanks.

_____ 1.

_____ 2.

_____ 3.

_____ 4.

_____ 5.

_____ 6.

_____ 7.

_____ 8.

_____ 9.

_____ 10.

_____ 11.

_____ 12.

_____ 13.

_____ 14.

Standing on your toes as in ballet is __(1)__ of the foot. Walking on your heels is __(2)__.

Winding up for a pitch (as in baseball) can properly be called __(3)__. To keep your seat when riding a horse, the tendency is to __(4)__ your thighs.

In running, the action at the hip joint is __(5)__ in reference to the leg moving forward and __(6)__ in reference to the leg in the posterior position. When kicking a football, the action at the knee is __(7)__. In climbing stairs, the hip and knee of the forward leg are both __(8)__. You have just touched your chin to your chest; this is __(9)__ of the neck.

Using a screwdriver with a straight arm requires __(10)__ of the arm. Consider all the movements of which the arm is capable. One often used for strengthening all the upper arm and shoulder muscles is __(11)__.

Moving the head to signify "no" is __(12)__. Action that moves the distal end of the radius across the ulna is __(13)__. Raising the arms laterally away from the body is called __(14)__ of the arms.

**16.** The terms provided in the key are often used to describe the manner in which muscles interact with other muscles. Select the key terms that apply to the following definitions and insert the correct letter or term in the answer blanks.

*Key Choices*

A. Antagonist       B. Fixator       C. Prime mover       D. Synergist

_____ 1. Agonist

_____ 2. Postural muscles for the most part

_____ 3. Stabilizes a joint so that the prime mover can act at more distal joints

_____ 4. Performs the same movement as the prime mover

_____ 5. Reverses and/or opposes the action of a prime mover

_____ 6. Immobilizes the origin of a prime mover

**17.** Several criteria are applied to the naming of muscles. These are provided in Column B. Identify which criteria pertain to the muscles listed in Column A and enter the correct letter(s) in the answer blank.

| Column A | Column B |
|---|---|
| _____ 1. Gluteus maximus | A. Action of the muscle |
| _____ 2. Adductor magnus | B. Shape of the muscle |
| _____ 3. Biceps femoris | C. Location of the muscle's origin and/or insertion |
| _____ 4. Abdominis transversus | D. Number of origins |
| _____ 5. Extensor carpi ulnaris | E. Location of muscle relative to a bone or body region |
| _____ 6. Trapezius | F. Direction in which the muscle fibers run relative to some imaginary line |
| _____ 7. Rectus femoris | |
| _____ 8. External oblique | G. Relative size of the muscle |

# GROSS ANATOMY OF THE SKELETAL MUSCLES

## Muscles of the Head

**18.** Identify the major muscles described in Column A by choosing a response from Column B. Enter the correct letter in the answer blank. Select a different color for each muscle described and color in the coding circle and corresponding muscle on Figure 6–6.

| Column A | Column B |
|---|---|
| ◯ _____ 1. Used in smiling | A. Buccinator |
| ◯ _____ 2. Used to suck in your cheeks | B. Frontalis |
| ◯ _____ 3. Used in winking | C. Masseter |
| ◯ _____ 4. Used to form the horizontal frown crease on the forehead | D. Orbicularis oculi |
| ◯ _____ 5. The "kissing" muscle | E. Orbicularis oris |
| ◯ _____ 6. Prime mover of jaw closure | F. Sternocleidomastoid |
| ◯ _____ 7. Synergist muscle for jaw closure | G. Temporalis |
| ◯ _____ 8. Prime mover of head flexion; a two-headed muscle | H. Trapezius |
| | I. Zygomaticus |

Zygomatic bone

**Figure 6-6**

# Muscles of the Trunk

**19.** Identify the anterior trunk muscles described in Column A by choosing a response from Column B. Enter the correct letter in the answer blank. Then, for each muscle description that has a color-coding circle, select a different color to color the coding circle and corresponding muscle on Figure 6–7.

**Column A**

○ _____ 1. The name means "straight muscle of the abdomen"

○ _____ 2. Prime mover for shoulder flexion and adduction

○ _____ 3. Prime mover for shoulder abduction

○ _____ 4. Part of the abdominal girdle; forms the external lateral walls of the abdomen

○ _____ 5. Acting alone, each muscle of this pair turns the head toward the opposite shoulder

_____ 6. and 7. Besides the two abdominal muscles (pairs) named above, two muscle pairs that help form the natural abdominal girdle

_____ 8. Deep muscles of the thorax that promote the inspiratory phase of breathing

_____ 9. An unpaired muscle that acts with the muscles named immediately above to accomplish inspiration

**Column B**

A. Deltoid

B. Diaphragm

C. External intercostal

D. External oblique

E. Internal intercostal

F. Internal oblique

G. Latissimus dorsi

H. Pectoralis major

I. Rectus abdominis

J. Sternocleidomastoid

K. Transversus abdominis

Clavicle

Sternum

Aponeurosis
normally
overlying
this muscle
has been
removed

**Figure 6–7**

**20.** Identify the posterior trunk muscles described in Column A by choosing a response from Column B. Enter the correct letter in the answer blank. Select a different color for each muscle description with a coding circle and color the coding circles and corresponding muscles on Figure 6–8.

**Column A**

⃝ _____ 1. Muscle that allows you to shrug your shoulders or extend your head

⃝ _____ 2. Muscle that adducts the shoulder and causes extension of the shoulder joint

⃝ _____ 3. Shoulder muscle that is the antagonist of the muscle just described

_____ 4. Prime mover of back extension; a deep composite muscle consisting of three columns

_____ 5. Large paired superficial muscle of the lower back

**Column B**

A. Deltoid

B. Erector spinae

C. External oblique

D. Gluteus maximus

E. Latissimus dorsi

F. Trapezius

**Figure 6–8**

# Muscles of the Hip, Thigh, and Leg

**21.** Identify the muscles described in Column A by choosing a response from Column B. Enter the correct letter in the answer blank. Select a different color for each muscle description provided with a color-coding circle, and use it to color the coding circles and corresponding muscles on Figure 6–9. Complete the illustration by labeling those muscles provided with leader lines.

**Column A**

_____ 1. Hip flexor, deep in pelvis; a composite of two muscles

○ _____ 2. Used to extend the hip when climbing stairs

○ _____ 3. "Toe dancer's" muscle; a two-bellied muscle of the calf

○ _____ 4. Inverts and dorsiflexes the foot

○ _____ 5. Muscle group that allows you to draw your legs to the midline of your body, as when standing at attention

○ _____ 6. Muscle group that extends the knee

○ _____ 7. Muscle group that extends the thigh and flexes the knee

○ _____ 8. Smaller hip muscle commonly used as an injection site

○ _____ 9. Muscle group of the lateral leg; plantar flex and evert the foot

○ _____ 10. Strap-like muscle that is a weak thigh flexor; the "tailor's muscle"

○ _____ 11. Like the two-bellied muscle that lies over it, this muscle is a plantar flexor

**Column B**

A. Adductors

B. Biceps femoris

C. Fibularis muscles

D. Gastrocnemius

E. Gluteus maximus

F. Gluteus medius

G. Hamstrings

H. Iliopsoas

I. Quadriceps

J. Rectus femoris

K. Sartorius

L. Semimembranosus

M. Semitendinosus

N. Soleus

O. Tibialis anterior

P. Vastus intermedius

Q. Vastus lateralis

R. Vastus medialis

**A**

**B**

**Figure 6–9**

# Muscles of the Arm and Forearm

**22.** Identify the muscles described in Column A by choosing a response from Column B. Enter the correct letter in the answer blank. Then select different colors for each muscle description provided with a color-coding circle and use them to color in the coding circles and corresponding muscles on Figure 6–10.

### Column A

○ _____    1. Wrist flexor that follows the ulna

○ _____    2. Muscle that extends the fingers

    _____    3. Muscle that flexes the fingers

○ _____    4. Muscle that allows you to bend (flex) the elbow

○ _____    5. Muscle that extends the elbow

○ _____    6. Powerful shoulder abductor, used to raise the arm overhead

### Column B

A. Biceps brachii

B. Deltoid

C. Extensor carpi radialis

D. Extensor digitorum

E. Flexor carpi ulnaris

F. Flexor digitorum superficialis

G. Triceps brachii

**Figure 6–10**

# General Body Muscle Review

**23.** Complete the following statements describing muscles. Insert the correct answers in the answer blanks.

_____ 1.

_____ 2.

_____ 3.

_____ 4.

_____ 5.

_____ 6.

_____ 7.

_____ 8.

_____ 9.

_____ 10.

_____ 11.

Three muscles— (1) , (2) , and (3) —are commonly used for intramuscular injections in adults.

The insertion tendon of the (4) group contains a large sesamoid bone, the patella.

The triceps surae insert in common into the (5) tendon.

The bulk of the tissue of a muscle tends to lie (6) to the part of the body it causes to move.

The extrinsic muscles of the hand originate on the (7) .

Most flexor muscles are located on the (8) aspect of the body; most extensors are located (9) . An exception to this generalization is the extensor-flexor musculature of the (10) .

The pectoralis major and deltoid muscles act synergistically to (11) the arm.

**24.** Circle the term that does not belong in each of the following groupings.

1. Vastus lateralis      Vastus medialis      Knee extension      Biceps femoris

2. Latissimus dorsi      Pectoralis major      Shoulder adduction      Antagonists

3. Buccinator      Frontalis      Masseter      Mastication      Temporalis

4. Vastus medialis      Rectus femoris      Iliacus      Origin on coxal bone

**25.** Identify the numbered muscles in Figure 6–11 by placing the numbers in the blanks next to the following muscle names. Then select a different color for each muscle provided with a color-coding circle and color the coding circle and corresponding muscle in Figure 6–11.

○ _____    1.   Orbicularis oris

○ _____    2.   Pectoralis major

○ _____    3.   External oblique

○ _____    4.   Sternocleidomastoid

○ _____    5.   Biceps brachii

○ _____    6.   Deltoid

○ _____    7.   Vastus lateralis

○ _____    8.   Frontalis

○ _____    9.   Rectus femoris

○ _____    10. Sartorius

○ _____    11. Gracilis

○ _____    12. Adductor group

○ _____    13. Fibularis longus

○ _____    14. Temporalis

○ _____    15. Orbicularis oculi

○ _____    16. Zygomaticus

○ _____    17. Masseter

○ _____    18. Vastus medialis

○ _____    19. Tibialis anterior

○ _____    20. Transversus abdominus

○ _____    21. Rectus abdominis

1
2
3
4

5
6

7

8

9

10

11
12

13

14

15

16

17

18
19
20

21

**Figure 6–11**

**26.** Identify each of the numbered muscles in Figure 6–12 by placing the numbers in the blanks next to the following muscle names. Then select different colors for each muscle and color the coding circles and corresponding muscles on Figure 6–12.

◯ _____    1.  Adductor muscle

◯ _____    2.  Gluteus maximus

◯ _____    3.  Gastrocnemius

◯ _____    4.  Latissimus dorsi

◯ _____    5.  Deltoid

◯ _____    6.  Semitendinosus

◯ _____    7.  Soleus

◯ _____    8.  Biceps femoris

◯ _____    9.  Triceps brachii

◯ _____    10. External oblique

◯ _____    11. Gluteus medius

◯ _____    12. Trapezius

**Figure 6–12**

# DEVELOPMENTAL ASPECTS OF THE MUSCULAR SYSTEM

**27.** Complete the following statements concerning the embryonic development of muscles and their functioning throughout life. Insert your answers in the answer blanks.

_____ 1.

_____ 2.

_____ 3.

_____ 4.

_____ 5.

_____ 6.

_____ 7.

_____ 8.

_____ 9.

_____ 10.

_____ 11.

_____ 12.

1. The first movement of the baby detected by the mother-to-be is called the __(1)__.

An important congenital muscular disease that results in the degeneration of the skeletal muscles by young adulthood is called __(2)__.

A baby's control over muscles progresses in a __(3)__ direction as well as a __(4)__ direction. In addition, __(5)__ muscular control (that is, waving of the arms) occurs before __(6)__ control (pincer grasp) does.

Muscles will ordinarily stay healthy if they are __(7)__ regularly; without normal stimulation they __(8)__.

__(9)__ is a disease of the muscles, which results from some problem with the stimulation of muscles by acetylcholine. The muscles become progressively weaker in this disease.

With age, our skeletal muscles decrease in mass; this leads to a decrease in body __(10)__ and in muscle __(11)__. Muscle tissue that is lost is replaced by noncontractile __(12)__ tissue.

## INCREDIBLE JOURNEY

### A Visualization Exercise for the Muscular System

*As you straddle this structure, you wonder what is happening.*

**28.** Where necessary, complete statements by inserting the missing words in the numbered spaces.

_____ 1.

1. On this incredible journey, you will be miniaturized and enter a skeletal muscle cell to observe the events that occur during muscle contraction. You prepare yourself by donning a wet suit and charging your ion detector. Then you climb into a syringe to prepare for injection. Your journey will begin when you see the gleaming connective tissue covering, the __(1)__ of a single muscle cell. Once injected, you monitor your descent through the epidermis and subcutaneous tissue. When you reach the muscle cell surface, you see that it is punctuated with pits at relatively

_____  2.

_____  3.

_____  4.

_____  5.

_____  6.

_____  7.

_____  8.

_____  9.

_____  10.

regular intervals. Looking into the darkness and off in the distance, you can see a leash of fibers ending close to a number of muscle cells. Since all of these fibers must be from the same motor neuron, this functional unit is obviously a __(2)__. You approach the fiber ending on your muscle cell and scrutinize the __(3)__ junction there. As you examine the junction, minute fluid droplets leave the nerve ending and attach to doughnut-shaped receptors on the muscle cell membrane. This substance released by the nerve ending must be __(4)__. Then, as a glow falls over the landscape, your ion detector indicates ions are disappearing from the muscle cell exterior and entering the muscle pits. The needle drops from high to low as the __(5)__ ions enter the pits from the watery fluid outside. You should have expected this, since these ions must enter to depolarize the muscle cells and start the __(6)__.

Next, you begin to explore one of the surface pits. As the muscle jerks into action, you topple deep into the pit. Sparkling electricity lights up the wall on all sides. You grasp for a handhold. Finally successful, you pull yourself laterally into the interior of the muscle cell and walk carefully along what seems to be a log. Then, once again, you notice an eerie glow as your ion detector reports that __(7)__ ions are entering the cytoplasm rapidly. The "log" you are walking on "comes to life" and begins to slide briskly in one direction. Unable to keep your balance, you fall. As you straddle this structure, you wonder what is happening. On all sides, cylindrical structures—such as the one you are astride—are moving past other similar but larger structures. Suddenly you remember, these are the __(8)__ myofilaments that slide past the __(9)__ myofilaments during muscle contraction.

Seconds later, the forward movement ends, and you begin to journey smoothly in the opposite direction. The ion detector now indicates low __(10)__ ion levels. Since you cannot ascend the smooth walls of one of the entry pits, you climb from one myofilament to another to reach the underside of the sarcolemma. Then you travel laterally to enter a pit close to the surface and climb out onto the cell surface. Your journey is completed, and you prepare to leave your host once again.

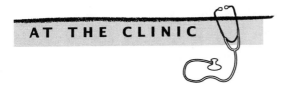

## AT THE CLINIC

**29.** Out of control during a temper tantrum, Malcolm smashed his fist through a glass door. Several tendons were sliced through at the anterior wrist. What movements are likely to be lost if tendon repair is not possible?

30. During an overambitious workout, a high school athlete pulls some muscles by forcing his knee into extension when his hip is already fully flexed. What muscles does he pull?

31. An emergency appendectomy is performed on Mr. Geiger. The incision was made at the lateral edge of the right iliac abdominopelvic region. Was his rectus abdominis cut?

32. Susan, a massage therapist, was giving Mr. Graves a backrub. What two broad superficial muscles of the back were receiving the "bulk" of her attention?

33. Mrs. Sanchez says that her 6-year-old son seems to be unusually clumsy and tires easily. The doctor notices that his calf muscles appear to be normal in size. If anything, they seem a bit enlarged rather than wasted. For what condition must the boy be checked? What is the prognosis?

34. People with chronic back pain occasionally get relief from a tummy tuck. How does this help?

35. Drugs are used to relax muscles during major surgery. Which of the two chemicals described next would make the best muscle relaxant and why? Chemical A binds with and blocks ACh receptors on muscle cells. Chemical B floods the sarcoplasm of muscle cells with calcium ions.

# *The Nervous System*

The nervous system is the master coordinating system of the body. Every thought, action, and sensation reflects its activity. Because of its complexity, the structures of the nervous system are described in terms of two principal divisions—the central nervous system (CNS) and the peripheral nervous system (PNS). The CNS, consisting of the brain and spinal cord, interprets incoming sensory information and issues instructions based on past experience. The PNS, consisting of cranial and spinal nerves and ganglia, provides the communication lines between the CNS and the body's muscles, glands, and sensory receptors. The nervous system is also divided functionally in terms of motor activities into the somatic and autonomic divisions. It is important, however, to recognize that these classifications are made for the sake of convenience and that the nervous system acts in an integrated manner both structurally and functionally.

Student activities provided in this chapter review neuron anatomy and physiology, identify the various structures of the central and peripheral nervous system, consider reflex and sensory physiology, and summarize autonomic nervous system anatomy and physiology. Because every body system is controlled, at least in part, by the nervous system, these understandings are extremely important to comprehending how the body functions as a whole.

**1.** List the three major functions of the nervous system.

1. _____

_____

2. _____

_____

3. _____

_____

# ORGANIZATION OF THE NERVOUS SYSTEM

**2.** Choose the key responses that best correspond to the descriptions provided in the following statements. Insert the appropriate letter or term in the answer blanks.

*Key Choices*

A. Autonomic nervous system
C. Peripheral nervous system (PNS)

B. Central nervous system (CNS)
D. Somatic nervous system

_____ 1. Nervous system subdivision that is composed of the brain and spinal cord

_____ 2. Subdivision of the PNS that controls voluntary activities such as the activation of skeletal muscles

_____ 3. Nervous system subdivision that is composed of the cranial and spinal nerves and ganglia

_____ 4. Subdivision of the PNS that regulates the activity of the heart and smooth muscle, and of glands; it is also called the involuntary nervous system

_____ 5. A major subdivision of the nervous system that interprets incoming information and issues orders

_____ 6. A major subdivision of the nervous system that serves as communication lines, linking all parts of the body to the CNS

# NERVOUS TISSUE—STRUCTURE AND FUNCTION

**3.** This exercise emphasizes the difference between neurons and neuroglia. Indicate which cell type is identified by the following descriptions. Insert the appropriate letter or term in the answer blanks.

*Key Choices*

A. Neurons
B. Neuroglia

_____ 1. Support, insulate, and protect cells

_____ 2. Demonstrate irritability and conductivity, and thus transmit electrical messages from one area of the body to another area

_____ 3. Release neurotransmitters

_____ 4. Are amitotic

_____ 5. Able to divide; therefore are responsible for most brain neoplasms

**4.** Relative to neuron anatomy, match the anatomical terms given in Column B
with the appropriate descriptions of functions provided in Column A. Place
the correct term or letter response in the answer blanks.

**Column A**          **Column B**

_____ 1. Releases neurotransmitters     A. Axon

_____ 2. Conducts electrical currents     B. Axonal terminal
                     toward the cell body
                                                       C. Dendrite

_____ 3. Increases the speed of impulse
                     transmission                      D. Myelin sheath

_____ 4. Location of the nucleus          E. Cell body

_____ 5. Generally conducts impulses
                     away from the cell body

**5.** Certain activities or sensations are listed below. Using key choices, select
the specific receptor type that would be activated by the activity or sensation
described. Insert the correct term(s) or letter response(s) in the answer blanks.
Note that more than one receptor type may be activated in some cases.

*Key Choices*

A. Bare nerve endings (pain)     C. Meissner's corpuscle     E. Pacinian corpuscle

B. Golgi tendon organ            D. Muscle spindle

**Activity or Sensation**                     **Receptor Type**

Walking on hot pavement                        1. (Identify two) _____

                                                  and _____

Feeling a pinch                                2. (Identify two) _____

                                                  and _____

Leaning on a shovel                            3. _____

Muscle sensations when rowing a boat           4. (Identify two) _____

                                                  and _____

Feeling a caress                               5. _____

**6.** Using key choices, select the terms identified in the following descriptions by inserting the appropriate letter or term in the spaces provided.

*Key Choices*

A. Afferent neuron          F. Neuroglia          K. Proprioceptors

B. Association neuron       G. Neurotransmitters  L. Schwann cells

C. Cutaneous sense organs   H. Nerve              M. Synapse

D. Efferent neuron          I. Nodes of Ranvier   N. Stimuli

E. Ganglion                 J. Nuclei             O. Tract

_____ 1. Sensory receptors found in the skin, which are specialized to detect temperature, pressure changes, and pain

_____ 2. Specialized cells that myelinate the fibers of neurons found in the PNS

_____ 3. Junction or point of close contact between neurons

_____ 4. Bundle of nerve processes inside the CNS

_____ 5. Neuron, serving as part of the conduction pathway between sensory and motor neurons

_____ 6. Gaps in a myelin sheath

_____ 7. Collection of nerve cell bodies found outside the CNS

_____ 8. Neuron that conducts impulses away from the CNS to muscles and glands

_____ 9. Sensory receptors found in muscle and tendons that detect their degree of stretch

_____ 10. Changes, occurring within or outside the body, that affect nervous system functioning

_____ 11. Neuron that conducts impulses toward the CNS from the body periphery

_____ 12. Chemicals released by neurons that stimulate other neurons, muscles, or glands

**7.** Figure 7–1 is a diagram of a neuron. First, label the parts indicated on the illustration by leader lines. Then choose different colors for each of the structures listed below and use them to color in the coding circles and corresponding structures in the illustration. Next, circle the term in the list of three terms to the left of the diagram that best describes this neuron's structural class. Finally, draw arrows on the figure to indicate the direction of impulse transmission along the neuron's membrane.

◯ Axon

◯ Dendrites

◯ Cell body

◯ Myelin sheath

**Unipolar**

**Bipolar**

**Multipolar**

**Figure 7–1**

**8.** List in order the *minimum* elements in a reflex arc from the stimulus to the activity of the effector. Place your responses in the answer blanks.

1. Stimulus

2. _____

3. _____

4. _____

5. Effector organ

9. Using the key choices, identify the terms defined in the following statements. Place the correct term or letter response in the answer blanks.

**Key Choices**

A. Action potential     D. Potassium ions     G. Sodium ions

B. Depolarization     E. Refractory period     H. Sodium-potassium pump

C. Polarized     F. Repolarization

_____ 1. Period of repolarization of the neuron during which it cannot respond to a second stimulus

_____ 2. State in which the resting potential is reversed as sodium ions rush into the neuron

_____ 3. Electrical condition of the plasma membrane of a resting neuron

_____ 4. Period during which potassium ions diffuse out of the neuron

_____ 5. Transmission of the depolarization wave along the neuron's membrane

_____ 6. The chief positive intracellular ion in a resting neuron

_____ 7. Process by which ATP is used to move sodium ions out of the cell and potassium ions back into the cell; completely restores the resting conditions of the neuron

10. Using the key choices, identify the types of reflexes involved in each of the following situations.

**Key Choices**

A. Somatic reflex(es)     B. Autonomic reflex(es)

_____ 1. Patellar (knee-jerk) reflex

_____ 2. Pupillary light reflex

_____ 3. Effectors are skeletal muscles

_____ 4. Effectors are smooth muscle and glands

_____ 5. Flexor reflex

_____ 6. Regulation of blood pressure

_____ 7. Salivary reflex

**11.** Refer to Figure 7–2, showing a reflex arc, as you complete this exercise. First, briefly answer the following questions by inserting your responses in the spaces provided.

1. What is the stimulus? _____

2. What tissue is the effector? _____

3. How many synapses occur in this reflex arc? _____

Next, select different colors for each of the following structures and use them to color in the coding circles and corresponding structures in the diagram. Finally, draw arrows on the figure indicating the direction of impulse transmission through this reflex pathway.

◯ Receptor region       ◯ Association neuron

◯ Afferent neuron       ◯ Efferent neuron

◯ Effector

**Figure 7–2**

**12.** Circle the term that does not belong in each of the following groupings.

1. Astrocytes      Neurons      Oligodendrocytes      Microglia

2. $K^+$ enters the cell      $K^+$ leaves the cell      Repolarization      Refractory period

3. Nodes of Ranvier      Myelin sheath      Unmyelinated      Saltatory conduction

4. Predictable response      Voluntary act      Involuntary act      Reflex

5. Oligodendrocytes      Schwann cells      Myelin      Microglia

6. Cutaneous receptors      Free dendritic endings      Stretch      Pain and touch

7. Cell interior      High $Na^+$      Low $Na^+$      High $K^+$

# CENTRAL NERVOUS SYSTEM
## Brain

**13.** Complete the following statements by inserting your answers in the answer blanks.

_____ 1.

_____ 2.

_____ 3.

_____ 4.

_____ 5.

The largest part of the human brain is the (paired) __(1)__. The other major subdivisions of the brain are the __(2)__ and the __(3)__. The cavities found in the brain are called __(4)__. They contain __(5)__.

**14.** Circle the terms indicating structures that are *not* part of the brain stem.

Cerebral hemispheres      Midbrain      Medulla

Pons      Cerebellum      Diencephalon

**15.** Complete the following statements by inserting your answers in the answer blanks.

_____ 1.

_____ 2.

_____ 3.

_____ 4.

_____ 5.

A __(1)__ is an elevated ridge of cerebral cortex tissue. The convolutions seen in the cerebrum are important because they increase the __(2)__. Gray matter is composed of __(3)__. White matter is composed of __(4)__, which provide for communication between different parts of the brain as well as with lower CNS centers. The lentiform nucleus, the caudate, and other nuclei are collectively called the __(5)__.

**16.** Figure 7–3 is a diagram of the right lateral view of the human brain. First, match the letters on the diagram with the following list of terms and insert the appropriate letters in the answer blanks. Then, select different colors for each of the areas of the brain provided with a color-coding circle and use them to color in the coding circles and corresponding structures in the diagram. If an identified area is part of a lobe, use the color you selected for the lobe but use *stripes* for that area.

_____ 1. ◯ Frontal lobe                  _____ 7.     Lateral sulcus

_____ 2. ◯ Parietal lobe                 _____ 8.     Central sulcus

_____ 3. ◯ Temporal lobe                 _____ 9. ◯ Cerebellum

_____ 4. ◯ Precentral gyrus              _____ 10. ◯ Medulla

_____ 5.     Parieto-occipital fissure   _____ 11. ◯ Occipital lobe

_____ 6. ◯ Postcentral gyrus             _____ 12. ◯ Pons

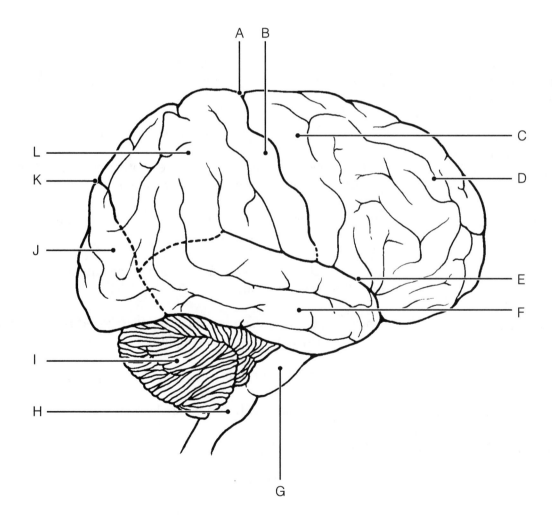

**Figure 7–3**

**17.** Figure 7–4 is a diagram of the sagittal view of the human brain. First, match the letters on the diagram with the following list of terms and insert the appropriate letter in each answer blank. Then, color the brain-stem areas blue and the areas where cerebrospinal fluid is found yellow.

_____ 1. Cerebellum

_____ 2. Cerebral aqueduct

_____ 3. Cerebral hemisphere

_____ 4. Cerebral peduncle

_____ 5. Choroid plexus

_____ 6. Corpora quadrigemina

_____ 7. Corpus callosum

_____ 8. Fourth ventricle

_____ 9. Hypothalamus

_____ 10. Medulla oblongata

_____ 11. Optic chiasma

_____ 12. Pineal body

_____ 13. Pituitary gland

_____ 14. Pons

_____ 15. Thalamus

**18.** Referring to the brain areas listed in Exercise 17, match the appropriate brain structures with the following descriptions. Insert the correct terms in the answer blanks.

_____ 1. Site of regulation of water balance and body temperature

_____ 2. Contains reflex centers involved in regulating respiratory rhythm in conjunction with lower brain-stem centers

_____ 3. Responsible for the regulation of posture and coordination of skeletal muscle movements

_____ 4. Important relay station for afferent fibers traveling to the sensory cortex for interpretation

_____ 5. Contains autonomic centers, which regulate blood pressure and respiratory rhythm, as well as coughing and sneezing centers

_____ 6. Large fiber tract connecting the cerebral hemispheres

_____ 7. Connects the third and fourth ventricles

_____ 8. Encloses the third ventricle

_____ 9. Forms the cerebrospinal fluid

_____ 10. Midbrain area that is largely fiber tracts; bulges anteriorly

_____ 11. Part of the limbic system; contains centers for many drives (rage, pleasure, hunger, sex, etc.)

**Figure 7-4**

**19.** For each of the following statements that is true, insert *T* in the answer blank.
If any of the statements are false, correct the underlined term by inserting
the correct term in the answer blank.

_____ 1. The primary sensory area of the cerebral hemisphere(s) is found
in the precentral gyrus.

_____ 2. Cortical areas involved in audition are found in the occipital lobe.

_____ 3. The primary motor area in the temporal lobe is involved in the
initiation of voluntary movements.

_____ 4. A specialized motor speech area located at the base of the precentral gyrus is called Wernicke's area.

_____ 5. The right cerebral hemisphere receives sensory input from the
right side of the body.

_____ 6. The pyramidal tract is the major descending voluntary motor tract.

_____ 7. Damage to the thalamus impairs consciousness and the
awake/sleep cycles.

→

_____ 8. A <u>flat</u> EEG is evidence of clinical death.

_____ 9. Beta waves are recorded when an individual is awake and <u>relaxed</u>.

## Protection of the CNS—Meninges and Cerebrospinal Fluid

**20.** Identify the meningeal (or associated) structures described here.

_____ 1. Outermost covering of the brain, composed of tough fibrous connective tissue

_____ 2. Innermost covering of the brain; delicate and vascular

_____ 3. Structures that return cerebrospinal fluid to the venous blood in the dural sinuses

_____ 4. Middle meningeal layer; like a cobweb in structure

_____ 5. Its outer layer forms the periosteum of the skull

**21.** Figure 7–5 shows a frontal view of the meninges of the brain at the level of the superior sagittal (dural) sinus. First, label the *arachnoid villi* on the figure. Then, select different colors for each of the following structures and use them to color the coding circles and corresponding structures in the diagram.

◯ Dura mater          ◯ Pia mater

◯ Arachnoid membrane      ◯ Subarachnoid space

Scalp

Bone of skull

Superior sagittal sinus

Gray matter of cerebral cortex

**Figure 7–5**

**22.** Complete the following statements by inserting your answers in the answer blanks.

_____ 1.

_____ 2.

_____ 3.

_____ 4.

_____ 5.

_____ 6.

_____ 7.

Cerebrospinal fluid is formed by capillary knots called __(1)__, which hang into the __(2)__ of the brain. Ordinarily, cerebro-spinal fluid flows from the lateral ventricles to the third ventricle and then through the __(3)__ to the fourth ventricle. Some of the fluid continues down the __(4)__ of the spinal cord, but most of it circulates into the __(5)__ by passing through three tiny openings in the walls of the __(6)__. As a rule, cerebrospinal fluid is formed and drained back into the venous blood at the same rate. If its drainage is blocked, a condition called __(7)__ occurs, which results in increased pressure on the brain.

# Brain Dysfunctions

**23.** Match the brain disorders listed in Column B with the conditions described in Column A. Place the correct answers in the answer blanks.

| Column A | Column B |
|---|---|
| _____ 1. Slight and transient brain injury | A. Alzheimer's disease |
| _____ 2. Traumatic injury that destroys brain tissue | B. Cerebral edema |
| _____ 3. Total nonresponsiveness to stimulation | C. Cerebrovascular accident (CVA) |
| _____ 4. May cause medulla oblongata to be wedged into foramen magnum by pressure of blood | D. Coma |
| _____ 5. After head injury, retention of water by brain | E. Concussion |
| _____ 6. Results when a brain region is deprived of blood or exposed to prolonged ischemia | F. Contusion |
| _____ 7. Progressive degeneration of the brain with abnormal protein deposits | G. Intracranial hemorrhage |
| _____ 8. Autoimmune disorder with extensive demyelination | H. Multiple sclerosis |
| _____ 9. A mini-stroke; fleeting symptoms of a CVA | I. Transient ischemic attack (TIA) |

# Spinal Cord

**24.** Complete the following statements by inserting your responses in the answer blanks.

_____ 1.

_____ 2.

_____ 3.

_____ 4.

_____ 5.

_____ 6.

_____ 7.

_____ 8.

_____ 9.

The spinal cord extends from the __(1)__ of the skull to the __(2)__ region of the vertebral column. The meninges, which cover the spinal cord, extend more inferiorly to form a sac from which cerebrospinal fluid can be withdrawn without damage to the spinal cord. This procedure is called a __(3)__. __(4)__ pairs of spinal nerves arise from the cord. Of these, __(5)__ pairs are cervical nerves, __(6)__ pairs are thoracic nerves, __(7)__ pairs are lumbar nerves, and __(8)__ pairs are sacral nerves. The tail-like collection of spinal nerves at the inferior end of the spinal cord is called the __(9)__.

**25.** Using key choices, select the appropriate terms to respond to the following descriptions referring to spinal cord anatomy. Place the correct term or letter in the answer blanks.

### Key Choices

A. Afferent (sensory)     C. Both afferent and efferent

B. Efferent (motor)       D. Association neurons (interneurons)

_____ 1. Neuron type found in the dorsal horn

_____ 2. Neuron type found in the ventral horn

_____ 3. Neuron type in a dorsal root ganglion

_____ 4. Fiber type in the ventral root

_____ 5. Fiber type in the dorsal root

_____ 6. Fiber type in a spinal nerve

**26.** Figure 7–6 is a cross-sectional view of the spinal cord. First, select different colors to identify the following structures and use them to color the coding circles and corresponding structures in the figure.

◯ Pia mater          ◯ Dura mater          ◯ Arachnoid

Then, identify the areas listed in the key choices by inserting the correct choices/ letter next to the appropriate leader line on the figure. Color the butterfly-shaped gray matter of the cord gray and the spinal roots and nerves yellow.

### Key Choices

A. Central canal          D. Dorsal root          G. Ventral horn

B. Column of white matter          E. Dorsal root ganglion          H. Ventral root

C. Dorsal horn          F. Spinal nerve

**Figure 7–6**

**27.** Using choices from Column B, indicate what would happen if the structures in Column A were damaged or transected. Place the correct letter in the answer blanks.

| Column A | Column B |
|---|---|
| _____ 1. Dorsal root of a spinal nerve | A. Loss of motor function |
| _____ 2. Ventral root of a spinal nerve | B. Loss of sensory function |
| _____ 3. Anterior ramus of a spinal nerve | C. Loss of both motor and sensory function |

# PERIPHERAL NERVOUS SYSTEM
## Structure of a Nerve

**28.** Figure 7–7 is a diagrammatic view of a nerve wrapped in its connective tissue coverings. Select different colors to identify the following structures and use them to color the coding circles and corresponding structures in the figure. Then, label each of the sheaths indicated by leader lines on the figure.

⭕ Endoneurium      ⭕ Perineurium      ⭕ Epineurium

Blood vessel

Fascicle

Myelin sheath     Axon

**Figure 7–7**

**29.** Complete the following statements by inserting your responses in the answer blanks.

_____ 1.

_____ 2.

_____ 3.

Another name for a bundle of nerve fibers is  (1) . Nerves carrying both sensory and motor fibers are called  (2)  nerves, whereas those carrying just sensory fibers are referred to as sensory, or  (3) , nerves.

**30.** Figure 7–8 is an anterior view of the principal nerves arising from the brachial plexus. Select five different colors and color the coding circles and the nerves listed below. Also, label each nerve by inserting its name at the appropriate leader line.

○ Axillary nerve

○ Musculocutaneous nerve

○ Median nerve

○ Radial nerve

○ Ulnar nerve

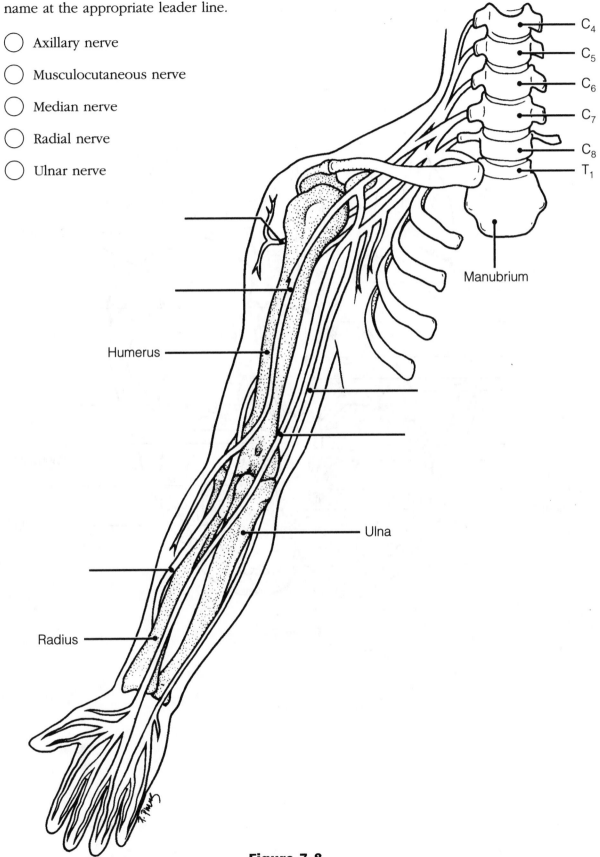

C₄
C₅
C₆
C₇
C₈
T₁

Manubrium

Humerus

Ulna

Radius

**Figure 7–8**

# Cranial Nerves

**31.** The 12 pairs of cranial nerves are indicated by leader lines in Figure 7–9. First, label each by name and Roman numeral on the figure and then color each nerve with a different color.

**Figure 7–9**

32. Provide the name and number of the cranial nerves involved in each of the following activities, sensations, or disorders. Insert your response in the answer blanks.

_____ 1. Shrugging the shoulders

_____ 2. Smelling a flower

_____ 3. Raising the eyelids and focusing the lens of the eye for accommodation; constriction of the eye pupils

_____ 4. Slows the heart; increases the mobility of the digestive tract

_____ 5. Involved in smiling

_____ 6. Involved in chewing food

_____ 7. Listening to music; seasickness

_____ 8. Secretion of saliva; tasting well-seasoned food

_____ 9. Involved in "rolling" the eyes (three nerves—provide numbers only)

_____ 10. Feeling a toothache

_____ 11. Reading *Tennis* magazine or this study guide

_____ 12. Purely sensory (three nerves—provide numbers only)

## Spinal Nerves and Nerve Plexuses

33. Complete the following statements by inserting your responses in the answer blanks.

_____ 1.

_____ 2.

_____ 3.

_____ 4.

The ventral rami of spinal nerves $C_1$ through $T_1$ and $L_1$ through $S_4$ take part in forming __(1)__, which serve the __(2)__ of the body. The ventral rami of $T_1$ through $T_{12}$ run between the ribs to serve the __(3)__. The posterior rami of the spinal nerves serve the __(4)__.

34. Name the major nerves that serve the following body areas. Insert your responses in the answer blanks.

_____ 1. Head, neck, shoulders (name plexus only)

_____ 2. Diaphragm

_____ 3. Posterior thigh

_____ 4. Leg and foot (name two)

→

_____ 5. Most anterior forearm muscles

_____ 6. Arm muscles

_____ 7. Abdominal wall (name plexus only)

_____ 8. Anterior thigh

_____ 9. Medial side of the hand

## Autonomic Nervous System

**35.** Identify, by color coding and coloring, the following structures in Figure 7–10, which depicts the major anatomical differences between the somatic and autonomic motor divisions of the PNS. Also identify by labeling all structures provided with leader lines.

◯ Somatic motor neuron

◯ ANS preganglionic neuron

◯ ANS ganglionic neuron

◯ Autonomic ganglion

◯ Gray matter of spinal cord (CNA)

◯ Effector of the somatic motor neuron

◯ Effector of the automatic motor neuron

◯ Myelin sheath

◯ White matter of spinal cord (CNS)

Small intestine (or other visceral organ)

CNS (spinal cord)

Skeletal muscle

**Figure 7–10**

**36.** The following table indicates a number of conditions. Use a check (✓) to show which division of the autonomic nervous system is involved in each condition.

| Condition | Sympathetic | Parasympathetic |
|---|---|---|
| 1. Postganglionic axons secrete norepinephrine; adrenergic fibers | | |
| 2. Postganglionic axons secrete acetylcholine; cholinergic fibers | | |
| 3. Long preganglionic axon, short postganglionic axon | | |
| 4. Short preganglionic axon, long postganglionic axon | | |
| 5. Arises from cranial and sacral nerves | | |
| 6. Arises from spinal nerves $T_1$ to $L_3$ | | |
| 7. Normally in control | | |
| 8. Fight-or-flight system | | |
| 9. Has more specific control | | |
| 10. Causes a dry mouth, dilates bronchioles | | |
| 11. Constricts eye pupils, decreases heart rate | | |

**37.** You are alone in your home late in the evening, and you hear an unfamiliar sound in your backyard. In the spaces provided, list four physiologic events promoted by the sympathetic nervous system that would help you to cope with this rather frightening situation.

1. _____

2. _____

3. _____

4. _____

# DEVELOPMENTAL ASPECTS OF THE NERVOUS SYSTEM

**38.** Complete the following statements by inserting your responses in the answer blanks.

| | |
|---|---|
| _____ 1. | Body temperature regulation is a problem in premature infants because the __(1)__ is not yet fully functional. Cerebral |
| _____ 2. | palsy involves crippling neuromuscular problems. It usually is a result of a lack of __(2)__ to the infant's brain during deliv- |
| _____ 3. | ery. Normal maturation of the nervous system occurs in a __(3)__ direction, and fine control occurs much later than __(4)__ |
| _____ 4. | muscle control. The sympathetic nervous system becomes less efficient as aging occurs, resulting in an inability to prevent |
| _____ 5. | sudden changes in __(5)__ when abrupt changes in position are made. The usual cause of decreasing efficiency of the ner- |
| _____ 6. | vous system as a whole is __(6)__. A change in intellect due to a gradual decrease in oxygen delivery to brain cells is called |
| _____ 7. | __(7)__. Death of brain neurons, which results from a sudden cessation of oxygen delivery, is called a __(8)__. |
| _____ 8. | |

## INCREDIBLE JOURNEY

### A Visualization Exercise for the Nervous System
*You climb on the first cranial nerve you see ...*

**39.** Where necessary, complete statements by inserting the missing words in the answer blanks.

_____ 1.  Nervous tissue is quite densely packed, and it is difficult to envision strolling through its various regions. Imagine instead that each of the various functional regions of the brain has a computerized room where you can observe what occurs in that particular area. Your assignment is to determine where you are at any given time during your journey through the nervous system.

You begin your journey after being miniaturized and injected into the warm pool of cerebrospinal fluid in your host's fourth ventricle. As you begin your stroll through the nervous tissue, you notice a huge area of branching white matter overhead. As you enter the first computer room you hear an announcement through the loudspeaker, "The pelvis is tipping too far posteriorly. Please correct. We are beginning to fall backward and will soon lose our balance." The computer responds imme- diately, decreasing impulses to the posterior hip muscles and increasing impulses to the anterior thigh muscles. "How is that, proprioceptor 1?" From this information, you determine that your first stop is the __(1)__.

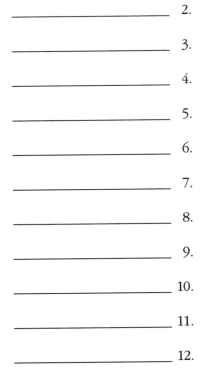

2.

3.

4.

5.

6.

7.

8.

9.

10.

11.

12.

2. At the next computer room, you hear, "Blood pressure to head is falling; increase sympathetic nervous system stimulation of the blood vessels." Then, as it becomes apparent that your host has not only stood up but is going to run, you hear, "Increase rate of impulses to the heart and respiratory muscles. We are going to need more oxygen and a faster blood flow to the skeletal muscles of the legs." You recognize that this second stop must be the  (2) .

Computer room 3 presents a problem. There is no loudspeaker here. Instead, incoming messages keep flashing across the wall, giving only bits and pieces of information. "Four hours since last meal: stimulate appetite center. Slight decrease in body temperature: initiate skin vasoconstriction. Mouth dry: stimulate thirst center. Oh, a stroke on the arm: stimulate pleasure center." Looking at what has been recorded here—appetite, temperature, thirst, and pleasure—you conclude that this has to be the  (3) .

Continuing your journey upward toward the higher brain centers, finally you are certain that you have reached the cerebral cortex. The first center you visit is quiet, like a library with millions of "encyclopedias" of facts and recordings of past input. You conclude that this must be the area where  (4)  are stored, and that you are in the  (5)  lobe. The next stop is close by. As you enter the computer center, you once again hear a loudspeaker: "Let's have the motor instructions to say tintinnabulation. Hurry, we don't want them to think we're tongue-tied." This area is obviously  (6) . Your final stop in the cerebral cortex is a very hectic center. Electrical impulses are traveling back and forth between giant neurons, sometimes in different directions and sometimes back and forth between a small number of neurons. Watching intently, you try to make some sense out of these interactions, and suddenly realize that this *is* what is happening here. The neurons *are* trying to make some sense out of something, and this helps you decide that this must be the brain area where  (7)  occurs in the  (8)  lobe.

You hurry out of this center and retrace your steps back to the cerebrospinal fluid, deciding en route to observe a cranial nerve. You decide to pick one randomly and follow it to the organ it serves. You climb on to the first cranial nerve you see and slide down past the throat. Picking up speed, you quickly pass the heart and lungs and see the stomach and small intestine coming up fast. A moment later you land on the stomach and now you know that this wandering nerve has to be the  (9) . As you look upward, you see that the nerve is traveling almost straight up and that you'll have to find an alternative route back to the cerebrospinal fluid. You begin to walk posteriorly until you find a spinal nerve, which you follow until you reach the vertebral column. You squeeze between two adjacent vertebrae to follow the nerve to the spinal cord. With your pocket knife you cut away the tough connective tissue covering the cord. Thinking that the  (10)  covering deserves its name, you finally manage to cut an opening large enough to get through and you return to the warm bath of cerebrospinal fluid that it encloses. At this point you are in the  (11) , and from here you swim upward until you get to the lower brain stem. Once there, it should be an easy task to find the holes leading into the  (12)  ventricle, where your journey began.

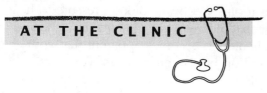

**40.** After surgery, patients are often temporarily unable to urinate, and bowel sounds are absent. Identify the division of the autonomic nervous system that is affected by anesthesia.

**41.** A brain tumor is found in a CT scan of Mr. Childs's head. The physician is assuming that it is not a secondary tumor (i.e., it did not spread from another part of the body) because an exhaustive workup has revealed no signs of cancer elsewhere in Mr. Childs's body. Is the brain tumor more likely to have developed from nerve tissue or from neuroglia? Why?

**42.** Amy, a high-strung teenager, was suddenly startled by a loud bang that sounded like a gunshot. Her heartbeat accelerated rapidly. When she realized that the noise was only a car backfiring, she felt greatly relieved but her heart kept beating heavily for several minutes more. Why does it take a long time to calm down after we are scared?

**43.** While working in the emergency room, you admit two patients who were in an auto collision. One is dead on arrival, having sustained a severed spinal cord at the level of $C_2$. The other patient suffered a similar injury but at the level of $C_6$ and is still alive. Explain briefly, in terms of the origin and function of the phrenic nerves, why one injury was fatal and the other was not.

**44.** You have been told that the superior and medial part of the right precentral gyrus of your patient's brain has been destroyed by a stroke. What part of the body is the patient unable to move? On which side, right or left?

**45.** *Application of knowledge:* You have been given all of the information
needed to identify the brain regions involved in the following situations.
See how well your nervous system has integrated this information, and
name the brain region (or condition) most likely to be involved in each
situation. Place your responses in the answer blanks.

1. Following a train accident, a man with an obvious head injury was
   observed stumbling about the scene. An inability to walk properly and
   a loss of balance were quite obvious. What brain region was injured?

   _____

2. An elderly woman is admitted to the hospital to have a gallbladder
   operation. While she is being cared for, the nurse notices that she has
   trouble initiating movement and has a strange "pill-rolling" tremor of
   her hands. What cerebral area is most likely involved?

   _____

3. A child is brought to the hospital with a high temperature. The doctor
   states that the child's meninges are inflamed. What name is given to this
   condition?

   _____

4. A young woman is brought into the emergency room with extremely
   dilated pupils. Her friends state that she has overdosed on cocaine.
   What cranial nerve is stimulated by the drug?

   _____

5. A young man has just received serious burns, resulting from standing
   with his back too close to a bonfire. He is muttering that he never felt
   the pain. Otherwise, he would have smothered the flames by rolling on
   the ground. What part of his CNS might be malfunctional?

   _____

6. An elderly gentleman has just suffered a stroke. He is able to understand
   verbal and written language, but when he tries to respond, his words
   are garbled. What cortical region has been damaged by the stroke?

   _____

7. A 12-year-old boy suddenly falls to the ground, having an epileptic
   seizure. He is rushed to the emergency room of the local hospital for
   medication. His follow-up care includes a recording of his brain waves
   to try to determine the area of the lesion. What is this procedure called?

   _____

**46.** Marie Nolin exhibits slow, tentative movements and a very unstable gait. Examination reveals she cannot touch her finger to her nose with eyes closed. What is the name of this condition and what part of her brain is damaged?

**47.** Which would be the more likely result of injury to the posterior side of the spinal cord only—paralysis or paresthesia (loss of sensory input)? Explain your answer.

# CHAPTER 8

# *Special Senses*

The body's sensory receptors react to stimuli or changes occurring both within the body and in the external environment. When triggered, these receptors send nerve impulses along afferent pathways to the brain for interpretation, thus, allowing the body to assess and adjust to changing conditions so that homeostasis may be maintained.

The minute receptors of general sensation that react to touch—pressure, pain, temperature changes, and muscle tension—are widely distributed in the body. These are considered in Chapter 7. In contrast, receptors of the special senses—sight, hearing, equilibrium, smell, and taste—tend to be localized and in many cases are quite complex. The structure and function of the special sense organs are the subjects of the student activities in this chapter.

## THE EYE AND VISION

1. Complete the following statements by inserting your responses in the answer blanks.

_____ 1.

_____ 2.

_____ 3.

_____ 4.

Attached to the eyes are the __(1)__ muscles that allow us to direct our eyes toward a moving object. The anterior aspect of each eye is protected by the __(2)__, which have eyelashes projecting from their edges. Closely associated with the lashes are oil-secreting glands called __(3)__ that help to lubricate the eyes. Inflammation of the mucosa lining the eyelids and covering the anterior part of the eyeball is called __(4)__.

2. Trace the pathway that the secretion of the lacrimal glands takes from the surface of the eye by assigning a number to each structure. (Note that #1 will be *closest* to the lacrimal gland.)

_____ 1. Lacrimal sac          _____ 3. Nasolacrimal duct

_____ 2. Nasal cavity          _____ 4. Lacrimal canals

**141**

**3.** Identify each of the eye muscles indicated by leader lines in Figure 8–1. Color code and color each muscle a different color. Then, in the blanks below, indicate the eye movement caused by each muscle.

◯ 1. Superior rectus _____    ◯ 4. Lateral rectus _____

◯ 2. Inferior rectus _____    ◯ 5. Medial rectus _____

◯ 3. Superior oblique _____    ◯ 6. Inferior oblique _____

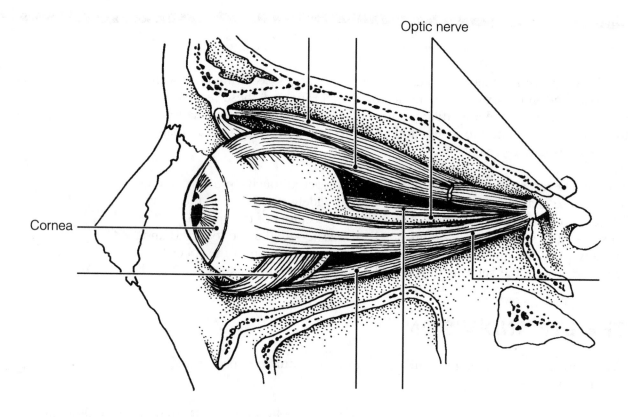

**Figure 8–1**

**4.** Three main accessory eye structures contribute to the formation of tears and/or aid in lubricating the eyeball. In the table, name each structure and then name its major secretory product. Indicate which of the secretions has antibacterial properties by circling that response.

| Accessory eye structures | Secretory product |
| --- | --- |
| 1. | |
| 2. | |
| 3. | |

**5.** Match the terms provided in Column B with the appropriate descriptions in Column A. Insert the correct letter response or corresponding term in the answer blanks.

**Column A**

_____ 1. Light bending

_____ 2. Ability to focus for close vision (under 20 feet)

_____ 3. Normal vision

_____ 4. Inability to focus well on close objects; farsightedness

_____ 5. Reflex-constriction of pupils when they are exposed to bright light

_____ 6. Clouding of the lens, resulting in loss of sight

_____ 7. Nearsightedness

_____ 8. Blurred vision, resulting from unequal curvatures of the lens or cornea

_____ 9. Condition of increasing pressure inside the eye, resulting from blocked drainage of aqueous humor

_____ 10. Medial movement of the eyes during focusing on close objects

_____ 11. Reflex constriction of the pupils when viewing close objects

_____ 12. Inability to see well in the dark; often a result of vitamin A deficiency

**Column B**

A. Accommodation

B. Accommodation pupillary reflex

C. Astigmatism

D. Cataract

E. Convergence

F. Emmetropia

G. Glaucoma

H. Hyperopia

I. Myopia

J. Night blindness

K. Photopupillary reflex

L. Refraction

**6.** The intrinsic eye muscles are under the control of which division of the nervous system? Circle the correct response.

1. Autonomic nervous system        2. Somatic nervous system

**7.** Complete the following statements by inserting your responses in the answer blanks.

_____ 1.

_____ 2.

_____ 3.

_____ 4.

_____ 5.        _____ 6.

A __(1)__ lens, like that of the eye, produces an image that is upside down and reversed from left to right. Such an image is called a __(2)__ image. In farsightedness, the light is focused __(3)__ the retina. The lens used to treat farsightedness is a __(4)__ lens. In nearsightedness, the light is focused __(5)__ the retina; it is corrected with a __(6)__ lens.

**8.** Using key choices, identify the parts of the eye described in the following statements. Insert the correct term or letter response in the answer blanks.

*Key Choices*

A. ◯ Aqueous humor       F. ◯ Fovea centralis       K. ◯ Sclera

B.   Canal of Schlemm     G. ◯ Iris                 L. ◯ Suspensory ligament

C. ◯ Choroid coat         H. ◯ Lens                 M. ◯ Vitreous humor

D. ◯ Ciliary body         I. ◯ Optic disk

E. ◯ Cornea               J. ◯ Retina

_____ 1. Attaches the lens to the ciliary body

_____ 2. Fluid that provides nutrients to the lens and cornea

_____ 3. The "white" of the eye

_____ 4. Area of retina that lacks photoreceptors

_____ 5. Contains muscle that controls the shape of the lens

_____ 6. Nutritive (vascular) tunic of the eye

_____ 7. Drains the aqueous humor of the eye

_____ 8. Tunic, containing the rods and cones

_____ 9. Gel-like substance that helps to reinforce the eyeball

_____ 10. Heavily pigmented tunic that prevents light scattering within the eye

_____ 11. _____ 12. Smooth muscle structures (intrinsic eye muscles)

_____ 13. Area of acute or discriminatory vision

_____ 14. _____ 15. Refractory media of the eye (#14–17)

_____ 16. _____ 17.

_____ 18. Anteriormost part of the sclera—your "window on the world"

_____ 19. Pigmented "diaphragm" of the eye

**9.** Using the key choice terms given in Exercise 8, identify the structures indicated by leader lines on the diagram of the eye in Figure 8–2. Select different colors for all structures provided with a color-coding circle in Exercise 8, and then use them to color the coding circles and corresponding structures in the figure.

**Figure 8–2**

**10.** In the following table circle the correct word under the vertical headings that describes events occurring within the eye during close and distant vision.

| Vision | Ciliary muscle | | Lens convexity | | Degree of light refraction | |
|---|---|---|---|---|---|---|
| 1. Distant | Relaxed | Contracted | Increased | Decreased | Increased | Decreased |
| 2. Close | Relaxed | Contracted | Increased | Decreased | Increased | Decreased |

**11.** Name in sequence the neural elements of the visual pathway, beginning with the retina and ending with the optic cortex.

Retina ⟶ _____ ⟶ _____ ⟶ _____

Synapse in thalamus ⟶ _____ ⟶ Optic cortex

**12.** Complete the following statements by inserting your responses in the answer blanks.

_____  1.   There are __(1)__ varieties of cones. One type responds most vigorously to __(2)__ light, another to __(3)__ light, and still
_____  2.   another to __(4)__ light. The ability to see intermediate colors such as purple results from the fact that more than one cone
_____  3.   type is being stimulated __(5)__. Lack of all color receptors results in __(6)__. Because this condition is sex linked, it occurs
_____  4.   more commonly in __(7)__. Black and white, or dim light, vision is a function of the __(8)__.

_____  5.

_____  6.   _____ 7.   _____ 8.

**13.** Circle the term that does not belong in each of the following groupings.

1. Choroid          Sclera          Vitreous humor          Retina

2. Ciliary body          Iris          Superior rectus          Choroid

3. Pupil constriction          Far vision          Accommodation          Bright light

4. Proprioceptors          Rods          Cones          Photoreceptors

5. Ciliary body          Iris          Suspensory ligaments          Lens

**14.** Complete the statements concerning rod photopigment and physiology by writing your responses in the answer blanks.

_____ 1.

_____ 2.

_____ 3.

_____ 4.

_____ 5.

_____ 6.

The bent or kinked form of retinal is combined with a protein called __(1)__ to form the visual pigment called __(2)__ . When light strikes the visual pigment, it straightens out and breaks down into its two components. This event is called __(3)__ because the purple color of the visual pigment changes to __(4)__ and finally becomes __(5)__ as retinal is converted all the way back to vitamin __(6)__ .

# THE EAR: HEARING AND BALANCE

**15.** Using key choices, select the terms that apply to the following descriptions. Place the correct letter in the answer blanks.

### Key Choices

A. Anvil (incus)          E. External auditory canal          I. Pinna          M. Tympanic membrane

B. Auditory tube          F. Hammer (malleus)          J. Round window          N. Vestibule

C. Cochlea          G. Oval window          K. Semicircular canals

D. Endolymph          H. Perilymph          L. Stirrup (stapes)

____ 1. ____ 2. ____ 3. Structures composing the outer ear

____ 4. ____ 5. ____ 6. Structures composing the bony or osseous labyrinth

____ 7. ____ 8. ____ 9. Collectively called the ossicles

____ 10. ____ 11. Ear structures not involved with hearing

_____ 12. Allows pressure in the middle ear to be equalized with the atmospheric pressure

_____ 13. Vibrates as sound waves hit it; transmits the vibrations to the ossicles

_____ 14. Contains the organ of Corti

_____ 15. Connects the nasopharynx and the middle ear

_____ 16. _____ 17. Contain receptors for the sense of equilibrium

_____ 18. Transmits the vibrations from the stirrup to the fluid in the inner ear

_____ 19. Fluid that bathes the sensory receptors of the inner ear

_____ 20. Fluid contained within the osseous labyrinth, which bathes the membranous labyrinth

**16.** Figure 8–3 is a diagram of the ear. Use anatomical terms (as needed) from key choices in Exercise 15 to correctly identify all structures in the figure provided with leader lines. Color all external ear structures yellow; color the ossicles red; color the equilibrium areas of the inner ear green; and color the inner ear structures involved with hearing blue.

**Figure 8–3**

**17.** Sound waves hitting the eardrum set it into vibration. Trace the pathway through which vibrations and fluid currents travel to finally stimulate the hair cells in the organ of Corti. Name the appropriate ear structures in their correct sequence and insert your responses in the answer blanks.

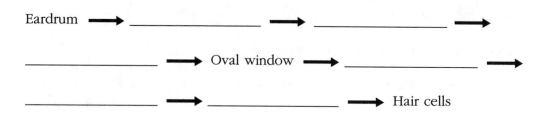

**18.** Figure 8–4 is a view of the structures of the membranous labyrinth. Correctly identify the following major areas of the labyrinth on the figure: *membranous semicircular canals, saccule* and *utricle,* and the *cochlear duct.* Next, correctly identify each of the receptor types shown in enlarged views (organ of Corti, crista ampullaris, and macula). Finally, using terms from the key choices below, identify all receptor structures provided with leader lines. (Some of these terms may need to be used more than once.)

**Figure 8–4**

**Key Choices**

A. Basilar membrane    E. Hair cells

B. Cochlear nerve    F. Otoliths

C. Cupula    G. Tectorial membrane

D. Gel    H. Vestibular nerve

**19.** Complete the following statements on the functioning of the static and dynamic equilibrium receptors by inserting the letter or term from the key choices in the answer blanks.

*Key Choices*

| | | |
|---|---|---|
| A. Angular/rotatory | E. Gravity | I.  Semicircular canals |
| B. Cupula | F. Perilymph | J.  Static |
| C. Dynamic | G. Proprioception | K. Utricle |
| D. Endolymph | H. Saccule | L. Vision |

_____  1.

_____  2.

_____  3.

_____  4.

_____  5.

_____  6.

_____  7.

_____  8.

_____  9.

_____  10.

_____  11.

The receptors for __(1)__ equilibrium are found in the crista ampullaris of the __(2)__. These receptors respond to changes in __(3)__ motion. When motion begins, the __(4)__ fluid lags behind and the __(5)__ is bent, which excites the hair cells. When the motion stops suddenly, the fluid flows in the opposite direction and again stimulates the hair cells. The receptors for __(6)__ equilibrium are found in the maculae of the __(7)__ and __(8)__. These receptors report the position of the head in space. Tiny stones found in a gel overlying the hair cells roll in response to the pull of __(9)__. As they roll, the gel moves and tugs on the hair cells, exciting them. Besides the equilibrium receptors of the inner ear, the senses of __(10)__ and __(11)__ are also important in helping to maintain equilibrium.

**20.** Indicate whether the following conditions relate to conduction deafness (*C*) or sensorineural (central) deafness (*S*). Place the correct letter choice in each answer blank.

_____  1. Can result from the fusion of the ossicles

_____  2. Can result from damage to the cochlear nerve

_____  3. Sound is heard in one ear but not in the other, during both bone and air conduction

_____  4. Often improved by a hearing aid

_____  5. Can result from otitis media

_____  6. Can result from excessive earwax or a perforated eardrum

_____  7. Can result from a blood clot in the auditory cortex of the brain

**21.** List three things about which a person with equilibrium problems might complain. Place your responses in the answer blanks.

_____ , _____ , and _____

**22.** Circle the term that does not belong in each of the following groupings.

1. Hammer          Anvil          Pinna          Stirrup

2. Tectorial membrane          Crista ampullaris          Semicircular canals          Cupula

3. Gravity          Angular motion          Sound waves          Rotation

4. Utricle          Saccule          Auditory tube          Vestibule

5. Vestibular nerve          Optic nerve          Cochlear nerve          Vestibulocochlear nerve

# CHEMICAL SENSES: SMELL AND TASTE

**23.** Complete the following statements by inserting your responses in the answer blanks.

_____ 1.

_____ 2.

_____ 3.

_____ 4.

_____ 5.

_____ 6.

_____ 7.

_____ 8.

_____ 9.

_____ 10.

_____ 11.

_____ 12.           15.

_____ 13.          _____ 16.

_____ 14.          _____ 17.

Three cranial nerves involved in transmitting impulses for the sense of taste are the __(1)__ , __(2)__ , and __(3)__ . Impulses for the sense of smell are transmitted by the __(4)__ nerve. The receptors for smell are located in the __(5)__ of the nasal passages; the act of __(6)__ increases the sensation, because it brings more air into contact with the receptors. The receptors for taste are found in clusterlike areas called __(7)__ , most of which are located on the sides of __(8)__ or __(9)__ papillae. The four basic taste sensations are __(10)__ , __(11)__ , __(12)__ , and __(13)__ . The most protective receptors are thought to be those that respond to __(14)__ substances. When nasal passages are congested, the sense of taste is decreased. This indicates that much of what is considered taste actually depends on the sense of __(15)__ . It is impossible to taste substances with a __(16)__ tongue, because foods must be dissolved (or in solution) to excite the taste receptors. The sense of smell is closely tied to the emotional centers of the brain (limbic region), and many odors bring back __(17)__ .

**24.** On Figure 8–5A, label the two types of tongue papillae containing taste buds. On Figure 8–5B, color the taste buds green. On Figure 8–5C, color the gustatory cells red, the supporting cells blue, and the cranial nerve fibers yellow. Add appropriate labels to the leader lines provided to identify the *taste pore* and *microvilli* of the gustatory cells.

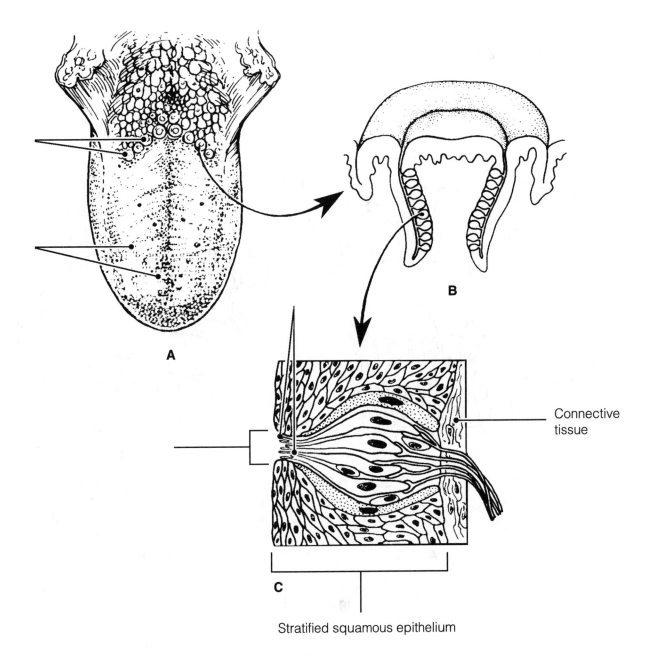

Connective tissue

Stratified squamous epithelium

**Figure 8–5**

**25.** Figure 8–6 illustrates the site of the olfactory epithelium in the nasal cavity (part A is an enlarged view of the olfactory receptor area). Select different colors to identify the structures listed below and use them to color the coding circles and corresponding structures in the illustration. Then add a label and leader line to identify the olfactory "hairs" and add arrows to indicate the direction of impulse transmission.

○ Olfactory neurons (receptor cells)    ○ Olfactory bulb

○ Supporting cells    ○ Cribriform plate of the ethmoid bone

○ Fibers of the olfactory tract    ○ Olfactory nerve filaments

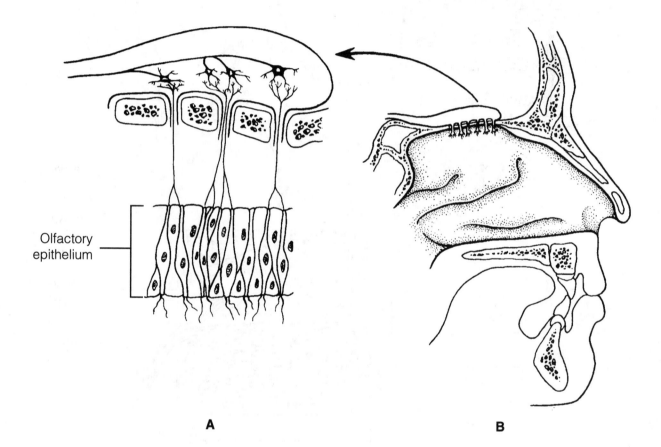

Olfactory
epithelium

A                                    B

**Figure 8–6**

**26.** Circle the term that does not belong in each of the following groupings.

1. Sweet        Musky        Sour        Bitter        Salty

2. Bipolar neuron        Epithelial cell        Olfactory receptor        Ciliated

3. Gustatory cell        Taste pore        Papillae        Neuron

4. Vagus nerve        Facial nerve        Glossopharyngeal nerve        Olfactory nerve

5. Olfactory receptor        High sensitivity        Variety of stimuli        Four receptor types

6. Sugars        Sweet        Saccharine        Metal ions        Amino acids


# DEVELOPMENTAL ASPECTS OF THE SPECIAL SENSES

**27.** Complete the following statements by inserting your responses in the answer blanks.

_____ 1.

_____ 2.

_____ 3.

_____ 4.

_____ 5.

_____ 6.

_____ 7.

_____ 8.

_____ 9.

The special sense organs are actually part of the __(1)__ and are formed very early in the embryo. Maternal infections, particularly __(2)__, may cause both deafness and __(3)__ in the developing child. Of the special senses, the sense of __(4)__ requires the most learning or takes longest to mature. All infants are __(5)__, but generally by school age emmetropic vision has been established. Beginning sometime after the age of 40, the eye lenses start to become less __(6)__ and cannot bend properly to refract the light. As a result, a condition of farsightedness, called __(7)__, begins to occur. __(8)__, a condition in which the lens becomes hazy or discolored, is a frequent cause of blindness. In old age, a gradual hearing loss, called __(9)__, occurs. A declining efficiency of the chemical senses is also common in the elderly.

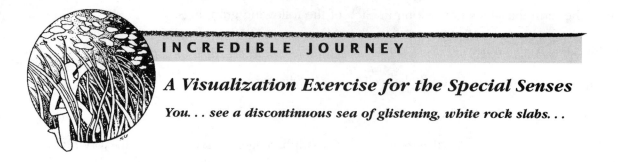

## A Visualization Exercise for the Special Senses

*You. . . see a discontinuous sea of glistening, white rock slabs. . .*

**28.** Where necessary, complete statements by inserting the missing words in the answer blanks.

_____ 1.

_____ 2.

_____ 3.

_____ 4.

_____ 5.

_____ 6.

_____ 7.

_____ 8.

_____ 9.

_____ 10.

_____ 11.

_____ 12.

Your present journey will take you through your host's inner ear to observe and document what you have learned about how hearing and equilibrium receptors work.

This is a very tightly planned excursion. Your host has been instructed to move his head at specific intervals and will be exposed to various sounds so that you can make certain observations. For this journey you are miniaturized and injected into the bony cavity of the inner ear, the __(1)__ , and are to make your way through its various chambers in a limited amount of time.

Your first observation is that you are in a warm sea of __(2)__ in the vestibule. To your right are two large sacs, the __(3)__ and __(4)__ . You swim over to one of these membranous sacs, cut a small semicircular opening in the wall, and wiggle through. Since you are able to see very little in the dim light, you set out to explore this area more fully. As you try to move, however, you find that your feet are embedded in a thick, gluelike substance. The best you can manage is slow-motion movements through this __(5)__ .

It is now time for your host's first scheduled head movement. Suddenly your world tips sharply sideways. You hear a roar (rather like an avalanche) and look up to see a discontinuous sea of glistening, white rock slabs sliding toward you. You protect yourself from these __(6)__ by ducking down between the hair cells that are bending vigorously with the motion of the rocks. Now that you have seen and can document the operation of a __(7)__ , a sense organ of __(8)__ equilibrium, you quickly back out through the hole you made.

Keeping in mind the schedule and the fact that it is nearly time for your host to be exposed to tuning forks, you swim quickly to the right, where you see what looks like the opening of a cave with tall seaweed waving gently in the current. Abruptly, as you enter the cave, you find that you are no longer in control of your movements, but instead are swept along in a smooth undulating pattern through the winding passageway of the cave, which you now know is the cavity of the __(9)__ . As you move up and down with the waves you see hair cells of the __(10)__ , the sense organ for __(11)__ , being vigorously disturbed below you. Flattening yourself against the chamber wall to prevent being carried further by the waves, you wait for the stimulus to stop. Meanwhile you are delighted by the electrical activity of the hair cells below you. As they depolarize to send impulses along the __(12)__ , the landscape appears to be alive with fireflies.

_____ 13.

_____ 14.

_____ 15.

_____ 16.

Now that you have witnessed the events for this particular sense receptor, you swim back through the vestibule toward your final observation area at the other end of the bony chambers. You recognize that your host is being stimulated again because of the change in fluid currents, but since you are not close to any of the sensory receptors, you are not sure just what the stimulus is. Then, just before you, three dark openings appear, the __(13)__ . You swim into the middle opening and see a strange structure that looks like the brush end of an artist's paintbrush; you swim upward and establish yourself on the soft brushy top portion. This must be the __(14)__ of the __(15)__ , the sensory receptor for __(16)__ equilibrium. As you rock back and forth in the gentle currents, a sudden wave of fluid hits you. Clinging to the hairs as the fluid thunders past you, you realize that there will soon be another such wave in the opposite direction. You decide that you have seen enough of the special senses and head back for the vestibule to leave your host once again.

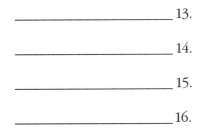

## AT THE CLINIC

**29.** An infant girl with strabismus is brought to the clinic. Tests show that she can control both eyes independently. What therapy will be tried before surgery?

**30.** A man in his early 60s comes to the clinic complaining of fuzzy vision. An eye examination reveals clouding of his lenses. What is his problem and what factors might have contributed to it?

**31.** Albinism is a condition in which melanin pigment is not made. How do you think vision is affected by albinism?

**32.** A man claiming to have difficulty seeing at night seeks help at the clinic. What is the technical name for this disorder? What dietary supplement will be recommended? If the condition has progressed too far, what retinal structures will be degenerated?

**33.** A child is brought to the speech therapist because she does not pronounce high-pitched sounds (like "s"). If it is determined that the spiral organ of Corti is the source of the problem, which region of the organ would be defective? Is this conduction or sensorineural deafness?

**34.** Little Biff's uncle tells the physician that 3-year-old Biff has frequent earaches and that a neighbor claims that Biff needs to have "ear tubes" put in. Upon questioning, the uncle reveals that Biff is taking swimming lessons and he can't remember the last time the boy had a sore throat. Does Biff have otitis media or otitis externa? Does he need ear tubes? Explain your reasoning.

**35.** Harry fell off a tall ladder and fractured the anterior cranial fossa of his skull. On arrival at the hospital, a watery, blood-tinged fluid was dripping from his right nostril. Several days later, Harry complained that he could no longer smell. What nerve was damaged in his fall?

**36.** Brian is brought to the clinic by his parents, who noticed that his right eye does not rotate laterally very well. The doctor explains that the nerve serving the lateral rectus muscle is not functioning properly. What nerve is he referring to?

**37.** When Mrs. Martinez visits her ophthalmologist, she complains of pain in her right eye. The intraocular pressure of that eye is found to be abnormally elevated. What is the name of Mrs. Martinez's probable condition? What causes it? What might be the outcome if the problem is not corrected?

# 9 The Endocrine System

The endocrine system, vital to homeostasis, plays an important role in regulating the activity of body cells. By acting through blood-borne chemical messengers, called hormones, the endocrine system organs orchestrate cellular changes that lead to growth and development, reproductive capability, and the physiological homeostasis of many body systems.

This chapter covers the location of the various endocrine organs in the body, the general function of the various hormones, and the consequences of their hypersecretion or hyposecretion.

## THE ENDOCRINE SYSTEM AND HORMONE FUNCTION—AN OVERVIEW

1. Complete the following statements by choosing answers from the key choices. Record the answers in the answer blanks.

*Key Choices*

A. Cardiovascular system     C. More rapid          E. Nervous system

B. Hormones                  D. Nerve impulses      F. Slower and more prolonged

_____ 1.

_____ 2.

_____ 3.

_____ 4.

_____ 5.

The endocrine system is a major controlling system in the body. Its means of control, however, is much __(1)__ than that of the __(2)__, the other major body system that acts to maintain homeostasis. Perhaps the reason for this is that the endocrine system uses chemical messengers, called __(3)__, instead of __(4)__. These chemical messengers enter the blood and are carried throughout the body by the activity of the __(5)__.

**2.** Complete the following statements by choosing answers from the key choices. Record the answers in the answer blanks.

*Key Choices*

A. Altering activity      F. Negative feedback      K. Steroid or amino acid–based

B. Anterior pituitary      G. Neural      L. Stimulating new or unusual activities

C. Hormonal      H. Neuroendocrine      M. Sugar or protein

D. Humoral      I. Receptors      N. Target cell(s)

E. Hypothalamus      J. Releasing hormones

_____ 1.

_____ 2.

_____ 3.

_____ 4.

_____ 5.

_____ 6.

_____ 7.

_____ 8.

_____ 9.

_____ 10.

_____ 11.

_____ 12.

_____ 13.

All cells do not respond to endocrine system stimulation. Only those that have the proper __(1)__ on their cell membranes are activated by the chemical messengers. These responsive cells are called the __(2)__ of the various endocrine glands. Hormones promote homeostasis by __(3)__ of body cells rather than by __(4)__. Most hormones are __(5)__ molecules.

The various endocrine glands are prodded to release their hormones by nerve fibers (a __(6)__ stimulus), by other hormones (a __(7)__ stimulus), or by the presence of increased or decreased levels of various other substances in the blood (a __(8)__ stimulus). The secretion of most hormones is regulated by a __(9)__ system, in which increasing levels of that particular hormone "turn off" its stimulus. The __(10)__ is called the master endocrine gland because it regulates so many other endocrine organs. However, it is in turn controlled by __(11)__ secreted by the __(12)__. The structure identified as #12 is also part of the brain, so it is appropriately called a __(13)__ organ.

**3.** For each key phrase, decide whether it better describes the mode of action of a steroid or amino acid–based hormone, and insert its key letter on the appropriate answer blank.

*Key Choices*

A. Binds to a plasma membrane receptor      D. Activates a gene to transcribe messenger RNA

B. Binds to a receptor in the cell's nucleus      E. Acts through a second messenger such as cyclic AMP

C. Is lipid soluble

Steroid hormones: _____      Amino acid–based hormones: _____

# THE MAJOR ENDOCRINE ORGANS

**4.** Figure 9–1 depicts the anatomical relationships between the hypothalamus and the anterior and posterior lobes of the pituitary in a highly simplified way. First, identify each of the structures listed below by color coding and coloring them on the diagram. Then, on the appropriate lines write in the names of the hormones that influence each of the target organs shown at the bottom of the diagram. Color the target organ diagrams as you like.

○ Hypothalamus                     ○ Anterior pituitary

○ Turk's saddle of the sphenoid bone      ○ Posterior pituitary

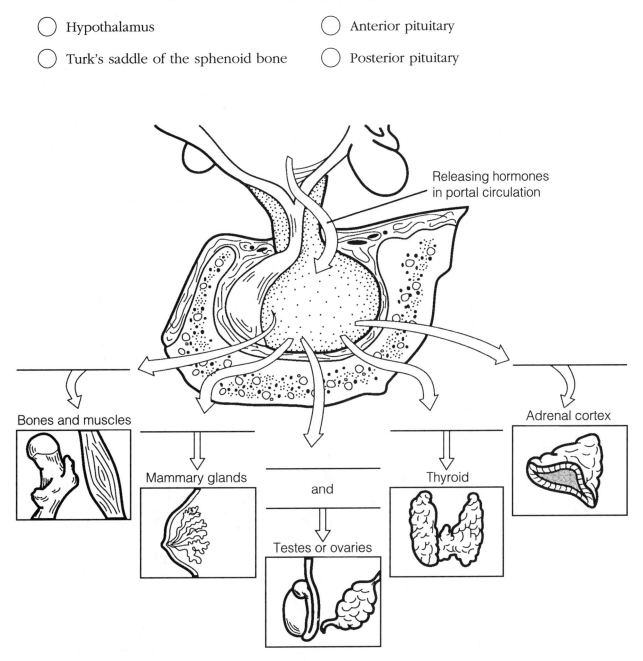

**Figure 9–1**

**5.** Figure 9–2 is a diagram of the various endocrine organs of the body. Next to each letter on the diagram, write the name of the endocrine-producing organ (or area). Then select different colors for each and color the corresponding organs in the illustration. To complete your identification of the hormone-producing organs, name the organs (not illustrated) described in items K and L.

K  Small glands that ride "horseback" on the thyroid

_____

L  Endocrine-producing organ present only in pregnant women

_____

**Figure 9–2**

**6.** For each of the following hormones, indicate the organ (or organ part)
producing or releasing the hormone by inserting the appropriate letters
from Figure 9–2 in the answer blanks.

_____ 1. ACTH

_____ 2. ADH

_____ 3. Aldosterone

_____ 4. Cortisone

_____ 5. Epinephrine

_____ 6. Estrogen

_____ 7. FSH

_____ 8. Glucagon

_____ 9. Insulin

_____ 10. LH

_____ 11. Melatonin

_____ 12. Oxytocin

_____ 13. Progesterone

_____ 14. Prolactin

_____ 15. PTH

_____ 16. Growth hormone

_____ 17. Testosterone

_____ 18. Thymosin

_____ 19. Thyrocalcitonin

_____ 20. Thyroxine

_____ 21. TSH

**7.** Name the hormone that best fits each of the following descriptions.
Insert your responses in the answer blanks.

_____ 1. Basal metabolic hormone

_____ 2. Programs T lymphocytes

_____ 3. Most important hormone regulating the amount of calcium
circulating in the blood; released when blood calcium levels drop

_____ 4. Helps to protect the body during long-term stressful situations
such as extended illness and surgery

_____ 5. Short-term stress hormone; aids in the fight-or-flight response;
increases blood pressure and heart rate, for example

_____ 6. Necessary if glucose is to be taken up by body cells

_____ 7. _____ 8.

_____ 9. _____ 10. Four tropic hormones

_____ 11. Acts antagonistically to insulin; produced by the same
endocrine organ

_____ 12. Hypothalamic hormone important in regulating water balance

_____ 13. _____ 14. Regulate the ovarian cycle

_____ 15. _____ 16. Directly regulate the menstrual
or uterine cycle

_____ 17. Adrenal cortex hormone involved in regulating salt levels
of body fluids

_____ 18. Necessary for milk production

**8.** Name the hormone that would be produced in inadequate amounts in the following conditions. Place your responses in the answer blanks.

_____ 1. Sexual immaturity

_____ 2. Tetany

_____ 3. Excessive urination without high blood glucose levels; causes dehydration and tremendous thirst

_____ 4. Goiter

_____ 5. Cretinism; a type of dwarfism in which the individual retains childlike proportions and is mentally retarded

_____ 6. Excessive thirst, high blood glucose levels, acidosis

_____ 7. Abnormally small stature, normal proportions

_____ 8. Miscarriage

_____ 9. Lethargy, falling hair, low basal metabolic rate, obesity (myxedema in the adult)

**9.** Name the hormone that would be produced in excessive amounts in the following conditions. Place your responses in the answer blanks.

_____ 1. Lantern jaw; large hands and feet (acromegaly in the adult)

_____ 2. Bulging eyeballs, nervousness, increased pulse rate, weight loss (Graves' disease)

_____ 3. Demineralization of bones; spontaneous fractures

_____ 4. Cushing's syndrome—moon face, depression of the immune system

_____ 5. Abnormally large stature, relatively normal body proportions

_____ 6. Abnormal hairiness; masculinization

**10.** List the cardinal symptoms of diabetes mellitus, and provide the rationale for the occurrence of each symptom.

1. _____

_____

2. _____

_____

3. _____

_____

**11.** The activity of many end organs is regulated by negative feedback. Figure 9–3A shows the basic elements of a homeostatic control system. Figure 9–3B shows a feedback loop with selected parts missing. Assume, for this system, that the stimulus that initiates it is declining $T_3$ and $T_4$ levels in the blood which produces a drop in metabolic rate. Fill in the information missing in the boxes to correctly complete this feedback loop. Also indicate whether it is a negative or positive feedback loop.

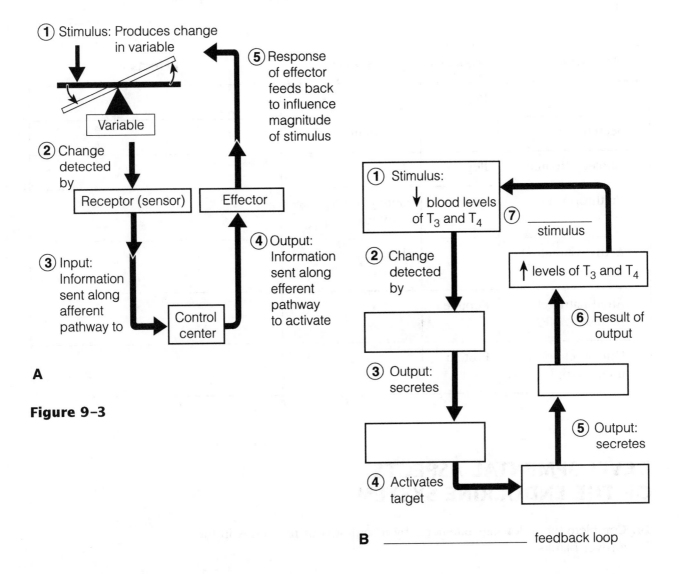

**Figure 9–3**

**12.** Circle the term that does not belong in each of the following groupings.

|     |                  |                              |                   |                           |
| --- | ---------------- | ---------------------------- | ----------------- | ------------------------- |
| 1.  | Posterior lobe   | Hormone storage              | Nervous tissue    | Anterior lobe             |
| 2.  | Steroid hormone  | Protein hormone              | Second messenger  | Membrane receptors        |
| 3.  | Catecholamines   | Norepinephrine               | Epinephrine       | Cortisol                  |
| 4.  | Calcitonin       | Increases blood $Ca^{2+}$    | Thyroid gland     | Enhances $Ca^{2+}$ deposit |
| 5.  | Glucocorticoids  | Steroids                     | Aldosterone       | Growth hormone            |

# OTHER HORMONE-PRODUCING TISSUES AND ORGANS

**13.** Besides the major endocrine organs, isolated clusters of cells produce hormones within body organs that are usually not associated with the endocrine system. A number of these hormones are listed in the table below. Complete the missing information on these hormones by filling in the blank spaces in the table.

| Hormone | Chemical makeup | Source | Effects |
|---|---|---|---|
| Gastrin | Peptide | | |
| Secretin | | Duodenum | |
| Cholecystokinin | Peptide | | |
| Erythropoietin | | Kidney in response to hypoxia | |
| Active vitamin $D_3$ | | Skin; activated by kidneys | |
| Atrial natriuretic peptide (ANP) | Peptide | | |
| Human chorionic gonadotropin (hCG) | Protein | | |

# DEVELOPMENTAL ASPECTS OF THE ENDOCRINE SYSTEM

**14.** Complete the following statements by inserting your responses in the answer blanks.

_____ 1.

_____ 2.

_____ 3.

_____ 4.

_____ 5.

_____ 6.

_____ 7.

Under ordinary conditions, the endocrine organs operate smoothly until old age. However, a __(1)__ in an endocrine organ may lead to __(2)__ of its hormones. A lack of __(3)__ in the diet may result in undersecretion of thyroxine. Later in life, a woman experiences a number of symptoms such as hot flashes and mood changes, which result from decreasing levels of __(4)__ in her system. This period of a woman's life is referred to as __(5)__, and it results in a loss of her ability to __(6)__. Because __(7)__ production tends to decrease in an aging person, adult-onset diabetes is common.

## A Visualization Exercise for the Endocrine System

*. . .you notice charged particles shooting pell-mell out of the bone matrix. . .*

**15.** Where necessary, complete statements by inserting the missing words in the answer blanks.

_____  1.

_____  2.

_____  3.

_____  4.

_____  5.

_____  6.

_____  7.

_____  8.

_____  9.

1. For this journey, you will be miniaturized and injected into a vein of your host. Throughout the journey, you will be traveling in the bloodstream. Your instructions are to record changes in blood composition as you float along, and to form some conclusions as to why they are occurring (that is, which hormone is being released).

Bobbing gently along in the slowly moving blood, you realize that there is a sugary taste to your environment; however, the sweetness begins to decrease quite rapidly. As the glucose levels of the blood have just decreased, obviously __(1)__ has been released by the __(2)__, so that the cells can take up glucose.

A short while later you notice that the depth of the blood in the vein you are traveling in has diminished substantially. To remedy this potentially serious situation, the __(3)__ will have to release more __(4)__, so the kidney tubules will reabsorb more water. Within a few minutes the blood becomes much deeper; you wonder if the body is psychic as well as wise.

As you circulate past the bones, you notice charged particles shooting pell-mell out of the bone matrix and jumping into the blood. You conclude that the __(5)__ glands have just released PTH, because the __(6)__ levels have increased in the blood. As you continue to move in the bloodstream, the blood suddenly becomes sticky sweet, indicating that your host must be nervous about something. Obviously, his __(7)__ has released __(8)__ to cause this sudden increase in blood glucose.

Sometime later, you become conscious of a humming activity around you, and you sense that the cells are very busy. Your host's __(9)__ levels appear to be sufficient, since his cells are certainly not sluggish in their metabolic activities. You record this observation and prepare to end this journey.

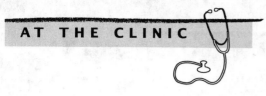

16. Pete is very short for his chronological age of 8. What physical features will allow you to determine quickly whether to check GH or thyroxine levels?

17. A young girl is brought to the clinic by her father. The girl fatigues easily and seems mentally sluggish. You notice a slight swelling in the anterior neck. What condition do you suspect? What are some possible causes and their treatments?

18. A 2-year-old boy is brought to the clinic by his anguished parents. He is developing sexually and shows an obsessive craving for salt. Blood tests reveal hyperglycemia. What endocrine gland is hypersecreting?

19. When the carnival came to a small town, the local health professionals and consumer groups joined forces to enforce truth-in-advertising laws to protect selected employees of the carnival. They demanded that the fat man, the dwarf, the giant, and the bearded lady be billed as "people with endocrine system problems" (which of course removed all the sensationalism usually associated with these attractions). Identify the endocrine disorder in each case and explain how (or why) the disorder produced the characteristic features of these four showpeople.

20. The brain is "informed" when we are stressed and the hypothalamus responds by secreting a releasing hormone called corticotropin-releasing hormone (CRH) which helps the body deal with the stressors. Outline this entire sequence, starting with CRH and ending with the release of cortisol. (Be sure to trace the hormone through the hypophyseal portal system and out of the pituitary gland.)

21. Mrs. Jackson claims she is not menstruating and reports that her breasts are producing milk, although she has never been pregnant. What hormone is being hypersecreted?

# *Blood*

Blood, the vital "life fluid" that courses through the body's blood vessels, provides the means by which the body's cells receive vital nutrients and oxygen and dispose of their metabolic wastes. As blood flows past the tissue cells, exchanges continually occur between the blood and the tissue cells, so that vital activities can go on continuously.

In this chapter, the student has an opportunity to review the general characteristics of whole blood and plasma, to identify the various formed elements (blood cells), and to recall their functions. Blood groups, transfusion reactions, clotting, and various types of blood abnormalities are also considered.

## COMPOSITION AND FUNCTIONS OF BLOOD

**1.** Complete the following description of the components of blood by writing the missing words in the answer blanks.

_____ 1.

_____ 2.

_____ 3.

_____ 4.

_____ 5.

_____ 6.

_____ 7.

_____ 8.

_____ 9.

_____ 10.

_____ 11.

In terms of its tissue classification, blood is classified as a __(1)__ because it has living blood cells, called __(2)__, suspended in a nonliving fluid matrix called __(3)__. The "fibers" of blood only become visible during __(4)__.

If a blood sample is centrifuged, the heavier blood cells become packed at the bottom of the tube. Most of this compacted cell mass is composed of __(5)__, and the volume of blood accounted for by these cells is referred to as the __(6)__. The less dense __(7)__ rises to the top and constitutes about 45% of the blood volume. The so-called "buffy coat" composed of __(8)__ and __(9)__ is found at the junction between the other two blood elements. The buffy coat accounts for less than __(10)__ % of blood volume.

Blood is scarlet red in color when it is loaded with __(11)__; otherwise, it tends to be dark red.

**2.** Using key choices, identify the cell type(s) or blood elements that fit the following descriptions. Insert the correct term or letter response in the spaces provided.

*Key Choices*

A. Red blood cell     D. Basophil     G. Lymphocyte

B. Megakaryocyte     E. Monocyte     H. Formed elements

C. Eosinophil     F. Neutrophil     I. Plasma

_____ 1. Most numerous leukocyte

_____ 2. _____ 3. _____ 4. Granular leukocytes

_____ 5. Also called an erythrocyte; anucleate

_____ 6. _____ 7. Actively phagocytic leukocytes

_____ 8. _____ 9. Agranular leukocytes

_____ 10. Fragments to form platelets

_____ 11. (A) through (G) are examples of these

_____ 12. Increases during allergy attacks

_____ 13. Releases histamine during inflammatory reactions

_____ 14. After originating in bone marrow, may be formed in lymphoid tissue

_____ 15. Contains hemoglobin

_____ 16. Primarily water, noncellular; the fluid matrix of blood

_____ 17. Increases in number during prolonged infections

_____ 18. Least numerous leukocyte

_____ 19. _____ 20. Also called white blood cells (#19–23)

_____ 21. _____ 22. _____ 23.

**3.** Figure 10–1 depicts in incomplete form the erythropoietin mechanism for regulating the rate of erythropoiesis. Complete the statements that have answer blanks, and then choose colors (other than yellow) for the color-coding circles and corresponding structures on the diagram. Color all arrows on the diagram yellow. Finally, indicate the normal life span of erythrocytes.

◯ Kidney      ◯ Red bone marrow      ◯ RBCs

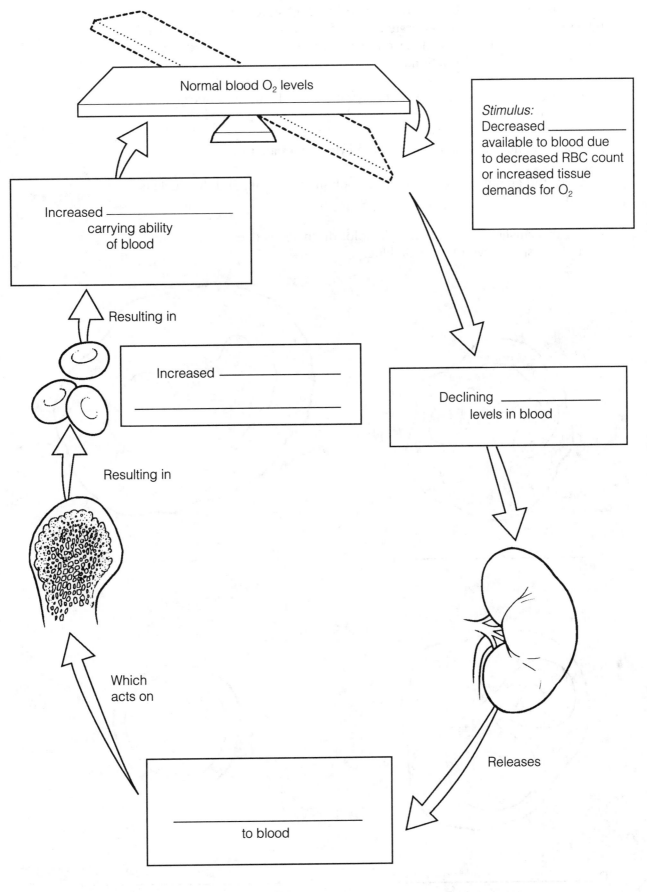

Normal blood O₂ levels

*Stimulus:*
Decreased _____
available to blood due
to decreased RBC count
or increased tissue
demands for O₂

Increased _____
carrying ability
of blood

Resulting in

Increased _____
_____

Declining _____
levels in blood

Resulting in

Which
acts on

Releases

_____
to blood

**Figure 10–1**

**4.** Four leukocytes are diagrammed in Figure 10–2. First, follow directions (given below) for coloring each leukocyte as it appears when stained with Wright's stain. Then, identify each leukocyte type by writing in the correct name in the blank below the illustration.

A. Color the granules pale violet, the cytoplasm pink, and the nucleus dark purple.

B. Color the nucleus deep blue and the cytoplasm pale blue.

C. Color the granules bright red, the cytoplasm pale pink, and the nucleus red/purple.

D. For this smallest white blood cell, color the nucleus deep purple/blue and the sparse cytoplasm pale blue.

A _____

B _____

C _____

D _____

**Figure 10–2**

**5.** For each true statement, insert *T*. If any of the statements are false, correct the underlined term by inserting the correction in the answer blank.

_____ 1. White blood cells (WBCs) move into and out of blood vessels by the process of chemotaxis.

_____ 2. An abnormal decrease in the number of white blood cells is leukopenia.

_____ 3. When blood becomes too acid or too basic, both the respiratory system and the liver may be called into action to restore it to its normal pH range.

_____ 4. The normal pH range of blood is 7.00 to 7.45.

_____ 5. The cardiovascular system of an average adult contains approximately 4 liters of blood.

_____ 6. Blood is circulated through the blood vessels by the pumping action of the heart.

_____ 7. An abnormal increase in the number of red blood cells is leukocytosis.

_____ 8. The normal RBC count is 3.5–4.5 million/mm$^3$.

_____ 9. Normal hemoglobin values are in the area of 42%–47% of the volume of whole blood.

_____ 10. An anemia resulting from a decreased RBC number causes the blood to become more viscous.

_____ 11. Phagocytic agranular WBCs are eosinophils.

_____ 12. The leukocytes particularly important in the immune response are monocytes.

**6.** Circle the term that does not belong in each of the following groupings.

1. Erythrocytes      Lymphocytes      Monocytes      Eosinophils

2. Neutrophils      Monocytes      Basophils      Eosinophils

3. Hemoglobin      Lymphocyte      Oxygen transport      Erythrocytes

4. Platelets      Monocytes      Phagocytosis      Neutrophils

5. Thrombus      Aneurysm      Embolus      Clot

6. Plasma      Nutrients      Hemoglobin      Wastes

7. Myeloid stem cell      Lymphocyte      Monocyte      Basophil

**7.** Check (✓) all the factors that would serve as stimuli for erythropoiesis.

_____ 1. Hemorrhage        _____ 3. Living at a high altitude

_____ 2. Aerobic exercise      _____ 4. Breathing pure oxygen

# HEMOSTASIS

**8.** Using key choices, correctly complete the following description of the blood-clotting process. Insert the key term or letter in the answer blanks.

*Key Choices*

A. Break          D. Fibrinogen         G. Serotonin

B. Erythrocytes     E. Platelets            H. Thrombin

C. Fibrin          F. Prothrombin        I. Thromboplastin

_____ 1.

_____ 2.

_____ 3.

_____ 4.

_____ 5.

_____ 6.

_____ 7.

_____ 8.

_____ 9.

Clotting begins when a __(1)__ occurs in a blood vessel wall. Almost immediately, __(2)__ cling to the blood vessel wall and release __(3)__, which helps to decrease blood loss by constricting the vessel. __(4)__ is also released by damaged cells in the area. This chemical substance causes __(5)__ to be converted to __(6)__. Once present, thrombin acts as an enzyme to attach __(7)__ molecules together to form long, threadlike strands of __(8)__, which then traps __(9)__ flowing by in the blood.

**9.** For each true statement, write *T.* If any statements are false, correct the underlined term by inserting the correction in the answer blank.

_____ 1. Normally, blood clots within 5–10 minutes.

_____ 2. The most important natural body anticoagulant is histamine.

_____ 3. Hemostasis means stoppage of blood flow.

# BLOOD GROUPS AND TRANSFUSIONS

**10.** Correctly complete the following table concerning ABO blood groups.

| Blood Type | Agglutinogens or antigens | Agglutinins or antibodies in plasma | Can donate blood to type | Can receive blood from type |
|---|---|---|---|---|
| 1. Type A | A | | | |
| 2. Type B | | anti-A | | |
| 3. Type AB | | | AB | |
| 4. Type O | none | | | |

**11.** What blood type is the *universal donor?* _____

The *universal recipient?* _____

**12.** When a person is given a transfusion of mismatched blood, a transfusion reaction occurs. Define the term "transfusion reaction" in the blanks provided here.

_____

_____

# DEVELOPMENTAL ASPECTS OF BLOOD

**13.** Complete the following statements by inserting your responses in the answer blanks.

_____ 1.

_____ 2.

_____ 3.

_____ 4.

A fetus has a special type of hemoglobin, hemoglobin __(1)__, that has a particularly high affinity for oxygen. After birth, the infant's fetal RBCs are rapidly destroyed and replaced by hemoglobin A–containing RBCs. When the immature infant liver cannot keep pace with the demands to rid the body of hemoglobin breakdown products, the infant's tissues become yellowed, or __(2)__.

Genetic factors lead to several congenital diseases concerning the blood. An anemia in which RBCs become sharp and "log-jam" in the blood vessels under conditions of low-oxygen tension in the blood is __(3)__ anemia. Bleeder's disease, or __(4)__, is a result of a deficiency of certain clotting factors.

→

_____ 5.

_____ 6.

_____ 7.

_____ 8.

_____ 9.

Diet is important to normal blood formation. Women are particularly prone to __(5)__ -deficiency anemia because of their monthly menses. A decreased efficiency of the gastric mucosa makes elderly individuals particularly susceptible to __(6)__ anemia as a result of a lack of intrinsic factor, which is necessary for vitamin __(7)__ absorption. An important problem in aged individuals is their tendency to form undesirable clots, or __(8)__. Both the young and the elderly are at risk for cancer of the blood, or __(9)__.

# INCREDIBLE JOURNEY

## A Visualization Exercise for the Blood

_Once inside, you quickly make a slash in the vessel lining . . ._

**14.** Where necessary, complete statements by inserting the missing words in the answer blanks.

_____ 1.

_____ 2.

_____ 3.

_____ 4.

_____ 5.

_____ 6.

_____ 7.

_____ 8.

_____ 9.

For this journey, you will be miniaturized and injected into the external iliac artery and will be guided by a fluorescent monitor into the bone marrow of the iliac bone. You will observe and report events of blood cell formation, also called __(1)__, seen there, and then move out of the bone into the circulation to initiate and observe the process of blood clotting, also called __(2)__. Once in the bone marrow, you watch as several large dark-nucleated stem cells, or __(3)__, begin to divide and produce daughter cells. To your right, the daughter cells eventually formed have tiny cytoplasmic granules and very peculiarly shaped nuclei that look like small masses of nuclear material connected by thin strands of nucleoplasm. You have just witnessed the formation of a type of white blood cell called the __(4)__. You describe its appearance and make a mental note to try to observe its activity later. Meanwhile you can tentatively report that this cell type functions as a __(5)__ to protect the body.

At another site, daughter cells arising from the division of a stem cell are difficult to identify initially. As you continue to observe the cells, you see that they, in turn, divide. Eventually some of their daughter cells eject their nuclei and flatten out to assume a disk shape. You assume that the kidneys must have released __(6)__, because those cells are __(7)__. That dark material filling their interior must be __(8)__, because those cells function to transport __(9)__ in the blood.

_____ 10.

_____ 11.

_____ 12.

_____ 13.

_____ 14.

_____ 15.

_____ 16.

_____ 17.

_____ 18.

_____ 19.

_____ 20.

_____ 21.

_____ 22.

_____ 23.

_____ 24.

_____ 25.

10. Now you turn your attention to the daughter cells being formed by the division of another stem cell. They are small round cells with relatively large round nuclei. In fact, their cytoplasm is very sparse. You record your observation of the formation of __(10)__ . They do not remain in the marrow very long after formation, but seem to enter the circulation almost as soon as they are produced. Some of those cells produce __(11)__ or act in other ways in the immune response. At this point, although you have yet to see the formation of __(12)__ , __(13)__ , __(14)__ , or __(15)__ , you decide to proceed into the circulation to make the blood-clotting observations.

You maneuver yourself into a small venule to enter the general circulation. Once inside, you quickly make a slash in the vessel lining, or __(16)__ . Almost immediately, what appear to be hundreds of jagged cell fragments swoop into the area and plaster themselves over the freshly made incision. You record that __(17)__ have just adhered to the damaged site. As you are writing, your chemical monitor flashes the message, "vasoconstrictor substance released." You record that __(18)__ has been released based on your observation that the vessel wall seems to be closing in. Peering out at the damaged site, you see that long ropelike strands are being formed at a rapid rate and are clinging to the site. You report that the __(19)__ mesh is forming and is beginning to trap RBCs to form the basis of the __(20)__ . Even though you do not have the equipment to monitor the intermediate steps of this process, you know that the platelets must have also released __(21)__ , which then converted __(22)__ to __(23)__ . This second enzyme then joined the soluble __(24)__ molecules together to form the network of strands you can see.

You carefully back away from the newly formed clot. You do _not_ want to disturb the area because you realize that if the clot detaches, it might become a life-threatening __(25)__ . Your mission here is completed, and you return to the entrance site.

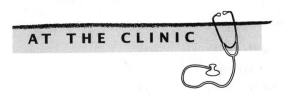

**AT THE CLINIC**

Correctly respond to five questions (#15–19) referring to the following situation. Mrs. Carlyle is pregnant for the first time. Her blood type is Rh negative, her husband is Rh positive, and their first child has been determined to be Rh positive. Ordinarily, the first such pregnancy causes no major problems, but baby Carlyle is born blue and cyanotic.

**15.** What is this condition, a result of Rh incompatibility, called?

16. Why is the baby cyanotic?

17. Because this is Mrs. Carlyle's first pregnancy, how can you account for the baby's problem?

18. Assume that baby Carlyle was born pink and healthy. What measures should be taken to prevent the previously described situation from happening in a second pregnancy with an Rh-positive baby?

19. Mrs. Carlyle's sister has had two miscarriages before seeking medical help with her third pregnancy. Blood typing shows that she, like her sister, is Rh negative; her husband is Rh positive. What course of treatment will be followed?

20. Ms. Pratt is claiming that Mr. X is the father of her child. Ms. Pratt's blood type is O negative. Her baby boy has type A positive blood. Mr. X's blood is typed and found to be B positive. Could he be the father of her child? _____ If not, what blood type would the father be expected to have? _____

21. Cancer patients being treated with chemotherapy drugs designed to destroy rapidly dividing cells are monitored closely for changes in their RBC and WBC counts. Why?

22. A red marrow biopsy is ordered for two patients—one a young child and the other an adult. The specimen is taken from the tibia of the child but from the iliac crest of the adult. Explain why different sites are used to obtain marrow samples in adults and children. (You might want to check Chapter 5 for this one.)

# 11 The Cardiovascular System

The major structures of the cardiovascular system, the heart and blood vessels, play a vital role in human physiology. The major function of the cardiovascular system is transportation. Using blood as the transport vehicle, the system carries nutrients, gases, wastes, antibodies, electrolytes, and many other substances to and from body cells. Its propulsive force is the contracting heart.

The anatomy and location of the heart and blood vessels and the important understandings of cardiovascular physiology (for example, cardiac cycle, ECG, and regulation of blood pressure) are the major topics of this chapter.

## CARDIOVASCULAR SYSTEM: THE HEART

**1.** Complete the following statements by inserting your answers in the answer blanks.

_____ 1.

_____ 2.

_____ 3.

_____ 4.

_____ 5.

_____ 6.

_____ 7.

_____ 8.

_____ 9.

_____ 10.

_____ 11.

The heart is a cone-shaped muscular organ located within the __(1)__ . Its apex rests on the __(2)__ , and its base is at the level of the __(3)__ rib. The coronary arteries that nourish the myocardium arise from the __(4)__ . The coronary sinus empties into the __(5)__ . Relative to the roles of the heart chambers, the __(6)__ are receiving chambers, whereas the __(7)__ are discharging chambers. The membrane that lines the heart and also forms the valve flaps is called the __(8)__ . The outermost layer of the heart is called the __(9)__ . The fluid that fills the pericardial sac acts to decrease __(10)__ during heart activity. The heart muscle, or myocardium, is composed of a specialized type of muscle tissue called __(11)__ .

**2.** The heart is called a double pump because it serves two circulations. Trace the flow of blood through both the pulmonary and systemic circulations by writing the missing terms in the answer blanks. Then, color regions transporting $O_2$-poor blood blue and regions transporting $O_2$-rich blood red on Figure 11–1. Finally, identify the various regions of the circulation shown in Figure 11–1 by labeling them using the key choices.

_____ 1.

_____ 2.

_____ 3.

_____ 4.

_____ 5.

_____ 6.

_____ 7.

_____ 8.

_____ 9.

_____ 10.

_____ 11.

_____ 12.

_____ 13.

From the right atrium through the tricuspid valve to the __(1)__, through the __(2)__ valve to the pulmonary trunk to the right and left __(3)__, to the capillary beds of the __(4)__, to the __(5)__, to the __(6)__ of the heart through the __(7)__ valve, to the __(8)__ through the __(9)__ semilunar valve, to the __(10)__, to the systemic arteries, to the __(11)__ of the body tissues, to the systemic veins, to the __(12)__ and __(13)__, which enter the right atrium of the heart.

### Key Choices

A. Vessels serving head and upper limbs

B. Vessels serving body trunk and lower limbs

C. Vessels serving the viscera

D. Pulmonary circulation

E. Pulmonary "pump"

F. Systemic "pump"

**Figure 11–1**

**3.** Figure 11–2 is an anterior view of the heart. Identify each numbered structure and write its name in the corresponding numbered space below the figure. Then, select different colors for each structure provided with a color-coding circle, and use them to color the coding circles and corresponding structures on the figure.

**Figure 11–2**

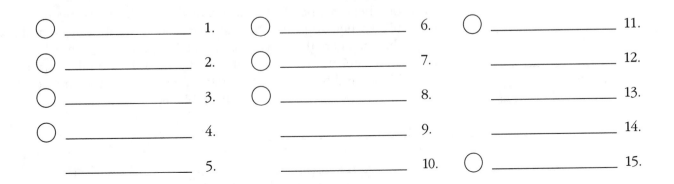

○ _____ 1.    ○ _____ 6.    ○ _____ 11.

○ _____ 2.    ○ _____ 7.    _____ 12.

○ _____ 3.    ○ _____ 8.    _____ 13.

○ _____ 4.    _____ 9.    _____ 14.

_____ 5.    _____ 10.    ○ _____ 15.

**4.** Figure 11–3 is a schematic drawing of the microscopic structure of cardiac muscle. Using different colors, color the coding circles of the structures listed below and the corresponding structures on the figure.

◯ Nuclei (with nucleoli)          ◯ Muscle fibers

◯ Intercalated discs          ◯ Striations

**Figure 11–3**

**5.** The events of one complete heartbeat are referred to as the cardiac cycle. Complete the following statements that describe these events. Insert your answers in the answer blanks.

_____ 1.

_____ 2.

_____ 3.

_____ 4.

_____ 5.

_____ 6.

The contraction of the ventricles is referred to as __(1)__, and the period of ventricular relaxation is called __(2)__. The monosyllables describing heart sounds during the cardiac cycle are __(3)__. The first heart sound is a result of closure of the __(4)__ valves; closure of the __(5)__ valves causes the second heart sound. The heart chambers that have just been filled when you hear the first heart sound are the __(6)__, and the chambers that have just emptied are the __(7)__. Immediately after the second heart sound, the __(8)__ are filling with blood, and the __(9)__ are empty. Abnormal heart sounds, or __(10)__, usually indicate valve problems.

_____ 7.    _____ 9.

_____ 8.    _____ 10.

**6.** Figure 11–4 is a diagram of the frontal section of the heart. Follow the instructions below to complete this exercise.

First, draw arrows to indicate the direction of blood flow through the heart. Draw the pathway of the oxygen-rich blood with red arrows, and trace the pathway of oxygen-poor blood with blue arrows.

Second, identify each of the elements of the intrinsic conduction system (numbers 1–5 on the figure) by inserting the appropriate terms in the blanks left of the figure. Then, indicate with green arrows the pathway that impulses take through this system.

Third, correctly identify each of the heart valves (numbers 6–9 on the figure) by inserting the appropriate terms in the blanks left of the figure, and draw in and identify by name the cordlike structures that anchor the flaps of the atrioventricular (AV) valves.

Fourth, use the numbers from the figure to identify the structures described below. Place the numbers in the lettered answer blanks.

_____ A. _____ B. Prevent backflow into the ventricles when the heart is relaxed

_____ C. _____ D. Prevent backflow into the atria when the ventricles are contracting

_____ E. AV valve with three flaps

_____ F. AV valve with two flaps

_____ G. The pacemaker of the intrinsic conduction system

_____ H. The point in the intrinsic conduction system where the impulse is temporarily delayed

_____ 1.

_____ 2.

_____ 3.

_____ 4.

_____ 5.

_____ 6.

_____ 7.

_____ 8.

_____ 9.

**Figure 11–4**

**7.** Match the terms provided in Column B with the statements given in Column A. Place the correct term or letter response in the answer blanks.

**Column A**

_____ 1. A recording of the electrical activity of the heart

_____ 2. The period during which the atria are depolarizing

_____ 3. The period during which the ventricles are repolarizing

_____ 4. The period during which the ventricles are depolarizing, which precedes their contraction

_____ 5. An abnormally slow heartbeat, that is, below 60 beats per minute

_____ 6. A condition in which the heart is uncoordinated and useless as a pump

_____ 7. An abnormally rapid heartbeat, that is, over 100 beats per minute

_____ 8. Damage to the AV node, totally or partially releasing the ventricles from the control of the sinoatrial (SA) node

_____ 9. Chest pain, resulting from ischemia of the myocardium

**Column B**

A. Angina pectoris

B. Bradycardia

C. Electrocardiogram

D. Fibrillation

E. Heart block

F. P wave

G. QRS wave

H. T wave

I. Tachycardia

**8.** A portion of an electrocardiogram is shown in Figure 11–5. On the figure identify the QRS complex, the P wave, and the T wave. Then, using a red pencil, bracket a portion of the recording equivalent to the length of one cardiac cycle. Using a blue pencil, bracket a portion of the recording in which the *ventricles* would be in diastole.

**Figure 11–5**

**9.** Complete the following statements relating to cardiac output by writing the missing terms in the answer blanks.

_____ 1.

_____ 2.

_____ 3.

_____ 4.

_____ 5.

_____ 6.

_____ 7.

_____ 8.

In the relationship CO = HR **x** SV, CO stands for __(1)__, HR stands for __(2)__, and SV stands for __(3)__. For the normal resting heart, the value of HR is __(4)__ and the value of SV is __(5)__. The normal average adult cardiac output, therefore, is __(6)__. The time for the entire blood supply to pass through the body is once each __(7)__.

According to Starling's law of the heart, the critical factor that determines force of heartbeat, or __(8)__, is the degree of __(9)__ of the cardiac muscle just before it contracts. Consequently, the force of heartbeat can be increased by increasing the amount of __(10)__ returned to the heart.

_____ 9. _____ 10.

**10.** Check (✓) all factors that lead to an *increase* in cardiac output by influencing either heart rate or stroke volume.

_____ 1. Epinephrine          _____ 6. Activation of the sympathetic nervous system

_____ 2. Thyroxine            _____ 7. Activation of the vagus nerves

_____ 3. Hemorrhage           _____ 8. Low blood pressure

_____ 4. Fear                 _____ 9. High blood pressure

_____ 5. Exercise             _____ 10. Fever

**11.** For each of the following statements that is true, write *T* in the answer blank. For any false statements, correct the underlined term by writing the correct term in the answer blank.

_____ 1. The resting heart rate is fastest in adult life.

_____ 2. Because the heart of the highly trained athlete hypertrophies, its stroke volume decreases.

_____ 3. If the right side of the heart fails, pulmonary congestion occurs.

_____ 4. In peripheral congestion, the feet, ankles, and fingers become edematous.

_____ 5. The pumping action of the healthy heart ordinarily maintains a balance between cardiac output and venous return.

**12.** Circle the term that does not belong in each of the following groupings.

1. Pulmonary trunk      Vena cava      Right side of heart      Left side of heart

2. QRS wave      T wave      P wave      Electrical activity of the ventricles

3. AV valves closed      AV valves opened      Ventricular systole      Semilunar valves open

4. Papillary muscles      Aortic semilunar valve      Tricuspid valve      Chordae tendineae

5. Tricuspid valve      Mitral valve      Bicuspid valve      Left AV valve

6. Ischemia      Infarct      Scar tissue repair      Heart block

# CARDIOVASCULAR SYSTEM: BLOOD VESSELS

**13.** Complete the following statements concerning blood vessels.

_____ 1.

_____ 2.

_____ 3.

_____ 4.

_____ 5.

_____ 6.

_____ 7.

The central cavity of a blood vessel is called the __(1)__. Reduction of the diameter of this cavity is called __(2)__, and enlargement of the vessel diameter is called __(3)__. Blood is carried to the heart by __(4)__ and away from the heart by __(5)__. Capillary beds are supplied by __(6)__ and drained by __(7)__.

**14.** Briefly explain in the space provided why valves are present in veins but not in arteries.

_____

_____

**15.** Name two events *occurring within the body* that aid in venous return. Place your responses in the blanks that follow.

_____ and _____

**16.** First, select different colors for each of the three blood vessel tunics listed in the key choices and illustrated in Figure 11–6 on p. 185. Color the color-coding circles and the corresponding structures in the three diagrams. In the blanks beneath the illustrations correctly identify each vessel type. In the additional spaces provided, list the structural details that allowed you to make the identifications. Then, using the key choices, identify the blood vessel tunics

described in each of the following descriptions. Insert the term or letter
of the key choice in the answer blanks.

**Key Choices**

A. ◯ Tunica intima      B. ◯ Tunica media      C. ◯ Tunica externa

_____ 1. Single thin layer of endothelium

_____ 2. Bulky middle coat, containing smooth muscle and elastin

_____ 3. Provides a smooth surface to decrease resistance to blood flow

_____ 4. The only tunic of capillaries

_____ 5. Also called the adventitia

_____ 6. The only tunic that plays an active role in blood pressure regulation

_____ 7. Supporting, protective coat

**Figure 11-6**

A _____      B _____      C _____

_____         _____         _____

_____         _____         _____

17. Figures 11–7 and 11–8 on pp. 186 and 187 illustrate the location of the most
    important arteries and veins of the body. The veins are shown in Figure 11–7.
    Color the veins blue and then identify each vein provided with a leader line on
    the figure. The arteries are shown in Figure 11–8. Color them red and then
    identify those indicated by leader lines on the figure. NOTE: If desired, the ves-
    sels identified may be colored differently to aid you in their later identification.

**Figure 11–7 Veins**

**Figure 11-8 Arteries**

**18.** Using key choices, identify the veins described as follows. Place the correct term or letter response in the answer blanks.

*Key Choices*

A. Anterior tibial    G. Common iliac    M. Hepatic portal    S. Radial

B. Azygos    H. Femoral    N. Inferior mesenteric    T. Renal

C. Basilic    I. Gastric    O. Inferior vena cava    U. Subclavian

D. Brachiocephalic    J. Gonadal    P. Internal iliac    V. Superior mesenteric

E. Cardiac    K. Great saphenous    Q. Internal jugular    W. Superior vena cava

F. Cephalic    L. Hepatic    R. Posterior tibial    X. Ulnar

_____ 1. _____ 2. Deep veins, draining the forearm

_____ 3. Vein that receives blood from the arm via the axillary vein

_____ 4. Veins that drain venous blood from the myocardium of the heart into the coronary sinus

_____ 5. Vein that drains the kidney

_____ 6. Vein that drains the dural sinuses of the brain

_____ 7. Two veins that join to become the superior vena cava

_____ 8. _____ 9. Veins that drain the leg and foot

_____ 10. Large vein that carries nutrient-rich blood from the digestive organs to the liver for processing

_____ 11. Superficial vein that drains the lateral aspect of the arm

_____ 12. Vein that drains the ovaries or testes

_____ 13. Vein that drains the thorax, empties into the superior vena cava

_____ 14. Largest vein below the thorax

_____ 15. Vein that drains the liver

_____ 16. _____ 17. _____ 18. Three veins that form/empty into the hepatic portal vein

_____ 19. Longest superficial vein of the body; found in the leg

_____ 20. Vein that is formed by the union of the external and internal iliac veins

_____ 21. Deep vein of the thigh

**19.** Figure 11–9 is a diagram of the hepatic portal circulation. Select different colors for the structures listed below and use them to color the color-coding circles and corresponding structures on the illustration.

⬤ Inferior mesenteric vein      ⬤ Splenic vein      ⬤ Hepatic portal vein

⬤ Superior mesenteric vein      ⬤ Gastric vein

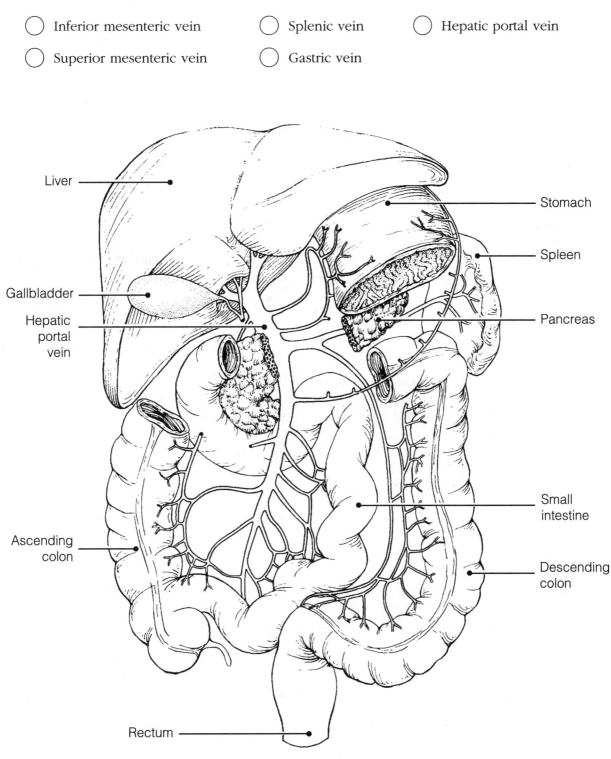

**Figure 11–9**

**20.** Using key choices, identify the *arteries* described as follows. Place the correct term or letter response in the spaces provided.

**Key Choices**

| | | | |
|---|---|---|---|
| A. Anterior tibial | H. Coronary | O. Intercostals | V. Renal |
| B. Aorta | I. Deep femoral | P. Internal carotid | W. Subclavian |
| C. Brachial | J. Dorsalis pedis | Q. Internal iliac | X. Superior mesenteric |
| D. Brachiocephalic | K. External carotid | R. Peroneal | Y. Vertebral |
| E. Celiac trunk | L. Femoral | S. Phrenic | Z. Ulnar |
| F. Common carotid | M. Hepatic | T. Posterior tibial | |
| G. Common iliac | N. Inferior mesenteric | U. Radial | |

_____ 1. _____ 2. Two arteries formed by the division of the brachiocephalic artery

_____ 3. First artery that branches off the ascending aorta; serves the heart

_____ 4. _____ 5. Two paired arteries, serving the brain

_____ 6. Largest artery of the body

_____ 7. Arterial network on the dorsum of the foot

_____ 8. Artery that serves the posterior thigh

_____ 9. Artery that supplies the diaphragm

_____ 10. Artery that splits to form the radial and ulnar arteries

_____ 11. Artery generally auscultated to determine blood pressure in the arm

_____ 12. Artery that supplies the last half of the large intestine

_____ 13. Artery that serves the pelvis

_____ 14. External iliac becomes this artery on entering the thigh

_____ 15. Major artery serving the arm

_____ 16. Artery that supplies most of the small intestine

_____ 17. The terminal branches of the dorsal, or descending, aorta

_____ 18. Arterial trunk that has three major branches, which serve the liver, spleen, and stomach

_____ 19. Major artery, serving the tissues external to the skull

_____ 20. _____ 21. _____ 22.
Three arteries, serving the leg inferior to the knee

_____ 23. Artery generally used to feel the pulse at the wrist

**21.** Figure 11–10 illustrates the arterial circulation of the brain. Select different colors for the following structures and use them to color the coding circles and corresponding structures in the diagram.

◯ Basilar artery                   ◯ Communicating branches

◯ Anterior cerebral arteries       ◯ Middle cerebral arteries

◯ Posterior cerebral arteries

**Figure 11–10**

22. Figure 11-11 illustrates the special fetal structures listed below. Select different colors for each and use them to color coding circles and corresponding structures in the diagram.

◯ Foramen ovale    ◯ Ductus arteriosus    ◯ Ductus venosus

◯ Umbilical arteries    ◯ Umbilical cord    ◯ Umbilical vein

Superior vena cava

Liver

Hepatic portal vein

Umbilicus

Inferior vena cava

Aorta

Common iliac artery

Internal iliac artery

**Figure 11-11**

Fetal bladder

**23.** Eight structures unique to the special circulations of the body are described here. Identify each, using the key choices. Place the correct terms or letters in the answer blanks.

*Key Choices*

A. Anterior cerebral artery     E. Ductus venosus     H. Posterior cerebral artery

B. Basilar artery     F. Foramen ovale     I. Umbilical artery

C. Circle of Willis     G. Middle cerebral artery     J. Umbilical vein

D. Ductus arteriosus

_____ 1. An anastomosis that allows communication between the posterior and anterior blood supplies of the brain

_____ 2. The vessel carrying oxygen and nutrient-rich blood to the fetus from the placenta

_____ 3. The shunt that allows most fetal blood to bypass the liver

_____ 4. Two pairs of arteries, arising from the internal carotid artery

_____ 5. The posterior cerebral arteries, serving the brain, arise from here

_____ 6. Fetal shunt between the aorta and pulmonary trunk that allows the lungs to be bypassed by the blood

_____ 7. Opening in the interatrial septum that shunts fetal blood from the right to the left atrium, thus bypassing the fetal lungs

**24.** Briefly explain in the space provided why the lungs are largely bypassed by the circulating blood in the fetus.

_____

_____

**25.** Circle the term that does not belong in each of the following groupings.

1. High pressure     Vein     Artery     Spurting blood

2. Carotid artery     Cardiac vein     Coronary sinus     Coronary artery

3. Increased venous return     Respiratory pump     Vasodilation     Milking action of skeletal muscles

4. High blood pressure     Hemorrhage     Weak pulse     Low cardiac output

5. Resistance     Friction     Vasodilation     Vasoconstriction

26. The following section relates to understandings concerning blood pressure and pulse. Match the items given in Column B with the appropriate descriptions provided in Column A. Place the correct term or letter response in the answer blanks.

| Column A | Column B |
|---|---|
| _____ 1. Expansion and recoil of an artery during heart activity | A. Over arteries |
| _____ 2. Pressure exerted by the blood against the blood vessel walls | B. Blood pressure |
| _____ 3. _____ 4. Factors related to blood pressure | C. Cardiac output |
| | D. Constriction of arterioles |
| _____ 5. Event primarily responsible for peripheral resistance | E. Diastolic blood pressure |
| _____ 6. Blood pressure during heart contraction | F. Peripheral resistance |
| _____ 7. Blood pressure during heart relaxation | G. Pressure points |
| _____ 8. Site where blood pressure determinations are normally made | H. Pulse |
| | I. Sounds of Korotkoff |
| _____ 9. Points at the body surface where the pulse may be felt | J. Systolic blood pressure |
| _____ 10. Sounds heard over a blood vessel when the vessel is partially compressed | K. Over veins |

27. Complete the following statements about capillary functions by placing answers from the key in the answer blanks. Use terms or letters from the key.

*Key Choices*

A. Blood          E. Fat soluble          H. Osmotic pressure

B. Capillary clefts     F. Hydrostatic pressure     I. Vesicles

C. Diffusion        G. Interstitial fluid       J. Water soluble

D. Fenestrations

_____ 1.  All exchanges to and from the blood and tissue cells occur through the __(1)__. Generally speaking, substances tend to
_____ 2.  move according to their concentration gradients by the process of __(2)__. Substances that are __(3)__ pass directly through
_____ 3.  the plasma membranes of the capillary endothelial cells; other

_____ 4.

_____ 5.

_____ 6.

_____ 7.

_____ 8.

_____ 9.

_____ 10.

_____ 11.

_____ 12.

_____ 13.

substances pass by means of or via __(4)__ , __(5)__ , or __(6)__ .
The most permeable capillaries are those exhibiting __(7)__ .
Capillaries that have __(8)__ and __(9)__ tend to be leaky and
two forces acting at capillary beds cause fluid flows. The
pressure that forces fluid out of the capillaries is __(10)__ and
the force that causes fluid to reenter the blood is __(11)__ .
Hence fluid is forced into the __(12)__ at the arteriole end
of the bed and into the __(13)__ at the venule end of the
capillary bed.

**28.** Indicate what effect the following factors have on blood pressure. Indicate
an increase in pressure by _I_ and a decrease in pressure by _D_. Place the
correct letter response in the answer blanks.

_____ 1. Increased diameter of the arterioles

_____ 2. Increased blood viscosity

_____ 3. Increased cardiac output

_____ 4. Increased pulse rate

_____ 5. Anxiety, fear

_____ 6. Increased urine output

_____ 7. Sudden change in position from
reclining to standing

_____ 8. Physical exercise

_____ 9. Physical training

_____ 10. Alcohol

_____ 11. Hemorrhage

_____ 12. Nicotine

_____ 13. Arteriosclerosis

**29.** Respond to the following exercise by placing brief answers in the spaces
provided. Assume someone has been injured in an automobile accident and
is bleeding profusely. What pressure point could you compress to help stop
the bleeding from the following areas?

_____ 1. Thigh

_____ 2. Forearm

_____ 3. Calf

_____ 4. Lower jaw

_____ 5. Thumb

**30.** For each of the following statements that is true, insert *T* in the answer blank. If any of the statements are false, correct the underlined term by inserting the correct word in the answer blank.

_____ 1. Renin, released by the kidneys, causes a <u>decrease</u> in blood pressure.

_____ 2. The decreasing efficiency of the sympathetic nervous system vasoconstrictor functioning, due to aging, leads to a type of hypotension called <u>sympathetic</u> hypotension.

_____ 3. Two body organs in which vasoconstriction rarely occurs are the heart and the <u>kidneys</u>.

_____ 4. A <u>sphygmomanometer</u> is used to take the apical pulse.

_____ 5. The pulmonary circulation is a <u>high</u>-pressure circulation.

_____ 6. The fetal equivalent of (functional) lungs and liver is the <u>placenta</u>.

_____ 7. Cold has a <u>vasodilating</u> effect.

_____ 8. <u>Thrombophlebitis</u> is called the silent killer.

# DEVELOPMENTAL ASPECTS OF THE CARDIOVASCULAR SYSTEM

**31.** Complete the following statements by inserting your responses in the answer blanks.

_____ 1.

_____ 2.

_____ 3.

_____ 4.

_____ 5.

_____ 6.

_____ 7.

_____ 8.

_____ 9.

_____ 10.

_____ 11.

The cardiovascular system forms early and the heart is acting as a functional pump by the __(1)__ week of development. The ductus arteriosus and foramen ovale allow the blood to bypass the nonfunctioning fetal __(2)__. Another fetal structure, the __(3)__, allows most of the blood to bypass the liver. The fetus is supplied with oxygen and nutrients via the __(4)__, which carries blood from the __(5)__ to the __(6)__. Metabolic wastes and carbon dioxide are removed from the fetus in blood carried by the __(7)__. These special bypass structures that exist to bypass the fetal lungs and liver become __(8)__ shortly after birth. Congenital heart defects (some resulting from the failure of the bypass structures to close) account for half of all infant __(9)__ resulting from congenital defects.

__(10)__ is a degenerative process that begins in youth but may take its toll in later life by promoting a myocardial infarct or stroke. Generally women have less atherosclerosis than men until after __(11)__, when estrogen production ends.

_____ 12.

_____ 13.

_____ 14.

_____ 15.

_____ 16.

Regular __(12)__ increases the efficiency of the cardiovascular system and helps to slow the progress of __(13)__ . A vascular problem that affects many in "standing professions" is __(14)__ . In this condition, the valves become incompetent, and the veins become twisted and enlarged, particularly in the __(15)__ and __(16)__ .

# INCREDIBLE JOURNEY

## A Visualization Exercise for the Cardiovascular System

**All about you are huge white cords, hanging limply from two flaps of endothelial tissue. . .**

**32.** Where necessary, complete the statements by inserting the missing word(s) in the answer blanks.

_____ 1.

_____ 2.

_____ 3.

_____ 4.

_____ 5.

Your journey starts in the pulmonary vein and includes a trip to part of the systemic circulation and a special circulation. You ready your equipment and prepare to be miniaturized and injected into your host.

Almost immediately after injection, you find yourself swept into a good-sized chamber, the __(1)__. However, you do not stop in this chamber, but continue to plunge downward into a larger chamber below. You land with a large splash and examine your surroundings. All about you are huge white cords, hanging limply from two flaps of endothelial tissue far above you. You report that you are sitting in the __(2)__ chamber of the heart, seeing the flaps of the __(3)__ valve above you. The valve is open and its anchoring cords, the __(4)__, are lax. Since this valve is open, you conclude that the heart is in the __(5)__ phase of the cardiac cycle.

→

_____ 6.

_____ 7.

_____ 8.

_____ 9.

_____ 10.

_____ 11.

_____ 12.

_____ 13.

_____ 14.

_____ 15.

_____ 16.

_____ 17.

_____ 18.

_____ 19.

_____ 20.

_____ 21.

6. Gradually you notice that the chamber walls seem to be closing in. You hear a thundering boom, and the whole chamber
7. vibrates as the valve slams shut above you. The cords, now
rigid and strained, form a cage about you, and you feel
8. extreme external pressure. Obviously, the heart is in a full-
fledged __(6)__. Then, high above on the right, the "roof"
9. opens, and you are forced through this __(7)__ valve. A
fraction of a second later, you hear another tremendous boom
10. that sends shock waves through the whole area. Out of the
corner of your eye, you see that the valve below you is
11. closed, and it looks rather like a pie cut into three wedges.

12. As you are swept along in this huge artery, the __(8)__, you
pass several branch-off points, but continue to career along,
13. straight down at a dizzying speed until you approach the
__(9)__ artery, feeding the small intestine. After entering this
14. artery and passing through successively smaller and smaller
subdivisions of it, you finally reach the capillary bed of the
15. small intestine. You watch with fascination as nutrient
molecules move into the blood through the single layer of
16. __(10)__ cells forming the capillary wall. As you move to the
opposite shore of the capillary bed, you enter a venule and
17. begin to move superiorly once again. The venules draining
the small intestine combine to form the __(11)__ vein, which in
18. turn combines with the __(12)__ vein to form the hepatic portal
vein that carries you into the liver. As you enter the liver, you
19. are amazed at the activity there. Six-sided hepatic cells,
responsible for storing glucose and making blood proteins,
20. are literally grabbing __(13)__ out of the blood as it percolates
slowly past them. Protective __(14)__ cells are removing bacte-
21. ria from the slowly moving blood. Leaving the liver through
the __(15)__ vein, you almost immediately enter the huge __(16)__,
which returns blood from the lower part of the body to the
__(17)__ of the heart. From here, you move consecutively through the right chambers of the heart
into the __(18)__. Soon that vessel splits and you are carried into a __(19)__ artery, which carries you to
the capillary beds of the __(20)__ and then back to the left side of the heart once again. After
traveling through the left side of the heart again, you leave your host when you are aspirated out
of the __(21)__ artery, which extends from the aorta to the axillary artery of the armpit.

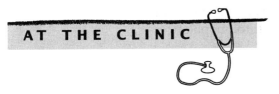

33. A man, en route to the hospital emergency room by ambulance, is in fibrillation. What is his cardiac output likely to be? He arrives at the emergency entrance DOA (dead on arrival). His autopsy reveals a blockage of the posterior interventricular artery. What is the cause of death?

34. Excessive vagal stimulation can be caused by severe depression. How would this be reflected in a routine physical examination?

35. A patient has swollen ankles and signs of degenerating organ functions. What is a likely diagnosis?

36. A routine scan of an elderly man reveals partial occlusion of the right internal carotid artery, yet blood supply to his cerebrum is unimpaired. What are two possible causes of the occlusion? What anastamosis is maintaining blood supply to the brain and by what (probable) route(s)?

37. A patient with a bone marrow cancer is polycythemic. Will his blood pressure be high or low? Why?

38. After a bout with bacterial endocarditis, scar tissue often stiffens the edges of the heart valves. How would this be picked up in a routine examination?

**39.** Len, an elderly man, is bedridden after a hip fracture. He complains of pain in his legs and thrombophlebitis is diagnosed. What is thrombophlebitis and what life-threatening complication can develop?

**40.** Mr. Langley is telling his friend about his recent visit to his doctor for a checkup. During his story, he mentions that the ECG revealed that he had a defective mitral valve and a heart murmur. Mr. Langley apparently misunderstood some of what the doctor explained to him about the diagnostic process. What has he misunderstood?

**41.** A less-than-respectable news tabloid announced that "Doctors show that exercise shortens life. Life expectancy is programmed into a set number of heartbeats; the faster your heart beats, the sooner you die." Even if this "theory" were true, what is wrong with the conclusion concerning exercise?

**42.** Mrs. Tuney says that when she stands up after lying down in the afternoon that she gets very dizzy. Her husband grumbles that "Its because she keeps the danged house too warm." He's right (in this particular case). Explain how this might cause her dizziness.

# 12 *The Lymphatic System and Body Defenses*

The lymphatic system, with its many lymphoid organs and vessels derived from veins of the cardiovascular system, is a rather strange system. Although both types of organs help to maintain homeostasis, these two elements of the lymphatic system have substantially different roles. The lymphatic vessels keep the cardiovascular system functional by maintaining blood volume. The lymphoid organs help defend the body from pathogens by providing operating sites for phagocytes and cells of the immune system.

The immune system, comprising the body's *specific defense system*, is a unique functional system made up of billions of individual cells, most of which are lymphocytes. The sole function of this defensive system is to protect the body against an incredible array of pathogens. In general, these "enemies" fall into three major camps: (1) microorganisms (bacteria, viruses, and fungi) that have gained entry into the body, (2) foreign tissue cells that have been transplanted (or, in the case of red blood cells, infused) into the body, and (3) the body's own cells that have become cancerous. The result of the immune system's activities is immunity, or specific resistance to disease.

The body is also protected by a number of nonspecific defenses provided by intact surface membranes such as skin and mucosae, and by a variety of cells and chemicals that can quickly mount an attack against foreign substances. The specific and nonspecific defenses enhance each other's effectiveness.

Chapter 12 tests your understanding of the functional roles of the various lymphatic system elements, and both the nonspecific and specific body defenses.

## THE LYMPHATIC SYSTEM
## Lymphatic Vessels

**1.** Complete the following statements by writing the missing terms in the answer blanks.

_____ 1.

_____ 2.

_____ 3.

_____ 4.

_____ 5.

_____ 6.

Although the cardiovascular system has a pump (the heart) and arteries, veins, and capillaries, the lymphatic system lacks two of these structures: the __(1)__ and __(2)__. Like the __(3)__ of the cardiovascular system, the vessels of the lymphatic system are equipped with __(4)__ to prevent backflow. The lymphatic vessels act primarily to pick up leaked fluid, now called __(5)__, and return it to the bloodstream. About __(6)__ of fluid is returned every 24 hours.

2. Figure 12–1 provides an overview of the lymphatic vessels. In part A, the relationship between lymphatic vessels and the blood vessels of the cardiovascular system is depicted schematically. Part B shows the different types of lymphatic vessels in a simple way. First, color code and color the following structures in Figure 12–1.

○ Heart          ○ Veins          ○ Lymphatic vessels/lymph node

○ Arteries       ○ Blood capillaries   ○ Loose connective tissue around
                                          blood and lymph capillaries

Then, identify by labeling these specific structures in part B:

A. Lymph capillaries      C. Lymphatic collecting vessels      E. Valves

B. Lymph duct             D. Lymph node                        F. Vein

Tissue fluid
(becomes
lymph)

A                                                              B

**Figure 12–1**

3. Circle the term that does not belong in each of the following groupings.

1. Blood capillary      Lymph capillary      Blind-ended      Permeable to proteins

2. Edema      Blockage of lymphatics      Elephantiasis      Inflammation

   Abundant supply of lymphatics

3. Skeletal muscle pump      Flow of lymph      Respiratory pump

   High-pressure gradient      Action of smooth muscle cells in walls of lymph vessels

4. Minivalves      Endothelial cell overlap      Impermeable      Lymphatic capillaries

# Lymph Nodes and Other Lymphoid Organs

**4.** Match the terms in Column B with the appropriate descriptions in Column A. More than one choice may apply in some cases.

<table>
<tr><td align="center">**Column A**</td><td align="center">**Column B**</td></tr>
<tr><td>_____ 1. The largest lymphatic organ; a blood reservoir</td><td>A. Lymph nodes</td></tr>
<tr><td></td><td>B. Peyer's patches</td></tr>
<tr><td>_____ 2. Filter lymph</td><td></td></tr>
<tr><td></td><td>C. Spleen</td></tr>
<tr><td>_____ 3. Particularly large and important during youth; produces hormones that help to program the immune system</td><td>D. Thymus</td></tr>
<tr><td></td><td>E. Tonsils</td></tr>
<tr><td>_____ 4. Collectively called MALT</td><td></td></tr>
<tr><td>_____ 5. Removes aged and defective red blood cells</td><td></td></tr>
<tr><td>_____ 6. Prevents bacteria from breaching the intestinal wall</td><td></td></tr>
</table>

**5.** Figure 12–2 depicts several different lymphoid organs. Label all lymphoid organs indicated by a leader line and add labels as necessary to identify the sites where the axillary, cervical, and inguinal lymph nodes would be located. Color the lymphoid organs as you like, and then shade in light green the portion of the body that is drained by the right lymphatic duct.

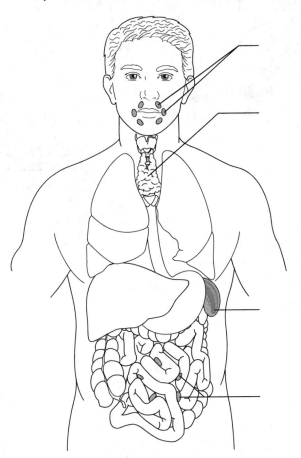

**Figure 12–2**

6. Figure 12–3 is a diagram of a lymph node. First, using the terms with color-coding circles, label all structures on the diagram that have leader lines. Color those structures as well. Then, add arrows to the diagram to show the direction of lymph flow through the organ. Circle the region that would approximately correspond to the medulla of the organ. Finally, answer the questions that follow.

○ Germinal centers of follicles

○ Cortex (other than germinal centers)

○ Medullary cords

○ Capsule and trabeculae

○ Hilus

○ Afferent lymphatics

○ Efferent lymphatics

○ Sinuses (subcapsular and medullary)

**Figure 12–3**

1. Which cell type is found in greatest abundance in the germinal centers?

   _____

2. What is the function of their daughter cells, the plasma cells?

   _____

3. What is the major cell type in cortical areas other than the germinal centers?

   _____

4. The third important cell type in lymph nodes (usually found clustered around

   the medullary sinuses) are the _____ .

   These cells act as _____ .

5. Of what importance is the fact that there are fewer efferent than afferent
   lymphatics associated with lymph nodes?

   _____

   _____

   _____

6. What structures assure the one-way flow of lymph through the node?

   _____

7. The largest collections of lymph nodes are found in what three body regions?

   _____

8. What is the general function of lymph nodes?

   _____

   _____

   _____

# BODY DEFENSES
## Nonspecific Body Defenses

7. The three major elements of the body's nonspecific defense system are: the

   (1) _____ , consisting of the skin and _____ ;

   defensive cells, such as (2) _____ and phagocytes; and a whole

   deluge of (3) _____ .

8. Indicate the sites of activity or the secretions of the nonspecific defenses by writing the correct terms in the answer blanks.

   1. Lysozyme is found in the body secretions called _____ and _____ .

   2. Fluids with an acid pH are found in the _____ and _____ .

   3. Sebum is a product of the _____ glands and acts at the surface

      of the _____ .

   4. Mucus is produced by mucus-secreting glands found in the respiratory and

      _____ system mucosae.

9. Circle the term that does not belong in each of the following groupings.

   1. Redness        Pain        Swelling        Itching        Heat

   2. Neutrophils        Macrophages        Phagocytes        Natural killer cells

   3. Inflammatory chemicals        Histamine        Kinins        Interferon

   4. Intact skin        Intact mucosae        Inflammation        First line of defense

   5. Interferons        Antiviral        Antibacterial        Proteins

**10.** Match the terms in Column B with the descriptions of the nonspecific defenses of the body in Column A. More than one choice may apply.

| Column A | Column B |
|---|---|
| _____ 1. Have antimicrobial activity | A. Acids |
| _____ 2. Provide mechanical barriers | B. Lysozyme |
| _____ 3. Provide chemical barriers | C. Mucosae |
| _____ 4. Entraps microorganisms entering the respiratory passages | D. Mucus |
|  | E. Protein-digesting enzymes |
| _____ 5. Part of the first line of defense | F. Sebum |
|  | G. Skin |

**11.** Describe the protective role of cilia in the respiratory tract. _____

_____

_____

**12.** Define *phagocytosis.* _____

_____

_____

**13.** Check (✓) all phrases that correctly describe the role of fever in body protection.

_____ 1. Is a normal response to pyrogens

_____ 2. Protects by denaturing tissue proteins

_____ 3. Reduces the availability of iron and zinc required for bacterial proliferation

_____ 4. Increases metabolic rate

**14.** List the three important accomplishments of the inflammatory response.

1. _____

2. _____

3. _____

**15.** Match the terms in Column B with the descriptions in Column A concerning events of the inflammatory response.

**Column A**

_____ 1. Accounts for redness and heat in an inflamed area

_____ 2. Inflammatory chemical released by injured cells

_____ 3. Promote release of white blood cells from the bone marrow

_____ 4. Cellular migration directed by a chemical gradient

_____ 5. Results from accumulation of fluid leaked from the bloodstream

_____ 6. Phagocytic offspring of monocytes

_____ 7. Leukocytes pass through the wall of a capillary

_____ 8. First phagocytes to migrate into the injured area

_____ 9. Walls off the area of injury

**Column B**

A. Chemotaxis

B. Diapedesis

C. Edema

D. Fibrin mesh

E. Histamine

F. Increased blood flow to an area

G. Inflammatory chemicals (including E)

H. Macrophages

I. Neutrophils

**16.** Complete the following description of the activation and activity of complement by writing the missing terms in the answer blanks.

_____ 1.

_____ 2.

_____ 3.

_____ 4.

_____ 5.

_____ 6.

Complement is a system of plasma __(1)__ that circulate in the blood in an inactive form. Complement is __(2)__ when it becomes attached to the surface of foreign cells (bacteria, fungi, red blood cells). One result of this complement fixation is that __(3)__ appear in the membrane of the foreign cell. This allows __(4)__ to rush in, which causes __(5)__ of the foreign cell. Some of the chemicals released during complement fixation enhance phagocytosis. This is called __(6)__. Others amplify the inflammatory response.

**17.** Describe the event that leads to the synthesis of interferon and the result
of its synthesis.

_____

_____

# Specific Body Defenses: The Immune System

*Antigens*

**18.** Complete the following statements relating to antigens by writing the
missing terms in the answer blanks.

_____ 1.

_____ 2.

_____ 3.

_____ 4.

Antigens are substances capable of mobilizing the __(1)__.
Of all the foreign molecules that act as complete antigens,
__(2)__ are the most potent. Small molecules are not usually
antigenic, but when they bind to self cell surface proteins
they may act as __(3)__, and then the complex is recognized
as foreign, or __(4)__.

*Cells of the Immune System: An Overview*

**19.** Using the key choices, select the term that correctly completes each
statement. Insert the appropriate term or letter in the answer blanks.

*Key Choices*

A. Antigen(s)      D. Cellular immunity      G. Lymph nodes

B. B cells          E. Humoral immunity      H. Macrophages

C. Blood            F. Lymph                 I. T cells

_____ 1.

_____ 2.

_____ 3.

_____ 4.

_____ 5.

_____ 6.

_____ 7.

_____ 8.      _____ 9.

Immunity is resistance to disease resulting from the presence
of foreign substances or __(1)__ in the body. When this
resistance is provided by antibodies released to body fluids,
the immunity is called __(2)__. When living cells provide the
protection, the immunity is referred to as __(3)__. The major
actors in the immune response are two lymphocyte
populations, the __(4)__ and the __(5)__. Phagocytic cells that act
as accessory cells in the immune response are the __(6)__.
Because pathogens are likely to use both __(7)__ and __(8)__ as
a means of getting around the body, __(9)__ and other
lymphatic tissues (which house the immune cells) are in an
excellent position to detect their presence.

**20.** A schematic of the life cycle of the lymphocytes involved in immunity is shown in Figure 12–4. First, select different colors for the areas listed below and use them to color the coding circles and the corresponding regions in the figure. If there is overlap, use stripes of a second color to indicate the second identification. Then respond to the statements following the figure, which relate to the two-phase differentiation process of B and T cells.

○ Area where immature lymphocytes arise

○ Area seeded by immunocompetent B and T cells

○ Area where T cells become immunocompetent

○ Area where the antigen challenge and clonal selection are likely to occur

○ Area where B cells become immunocompetent

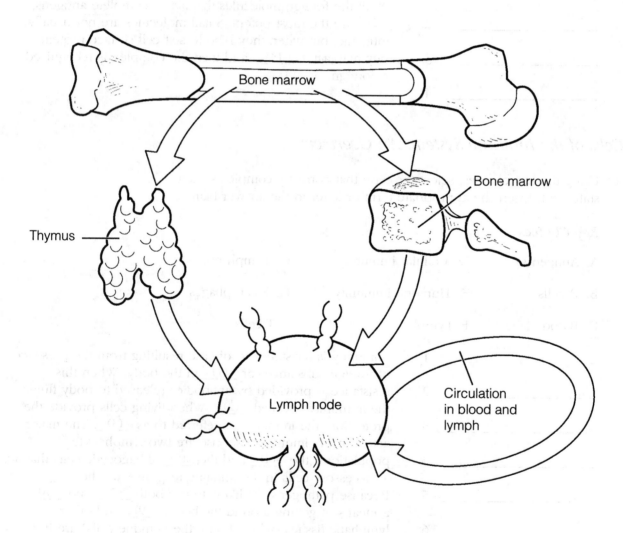

**Figure 12–4**

1. What signifies that a lymphocyte has become immunocompetent?

   _____

2. During what period of life does immunocompetence develop?

   _____

3. What determines which antigen a particular T or B cell will be able to recognize? (circle one)

   its genes          "its" antigen

4. What triggers the process of clonal selection in a T or B cell? (circle one)

   its genes          binding to "its" antigen

5. During development of immunocompetence, the ability to tolerate _____ must also occur if the immune system is to function normally.

**21.** T cells and B cells exhibit certain similarities and differences. Check (✓) the appropriate spaces in the table below to indicate the lymphocyte type that exhibits each characteristic.

| Characteristic | T cell | B cell |
|---|---|---|
| Originates in bone marrow from stem cells called hemocytoblasts | | |
| Progeny are plasma cells | | |
| Progeny include suppressors, helpers, and killers | | |
| Progeny include memory cells | | |
| Is responsible for directly attacking foreign cells or virus-infected cells | | |
| Produces antibodies that are released to body fluids | | |
| Bears a cell-surface receptor capable of recognizing a specific antigen | | |
| Forms clones upon stimulation | | |
| Accounts for most of the lymphocytes in the circulation | | |

**22.** Circle the term that does not belong in each of the following groupings.

1. Antibodies      Gamma globulin      Lymphokines      Immunoglobulins

2. Protein      Complete antigen      Nucleic acid      Hapten

3. Lymph nodes      Liver      Spleen      Thymus      Bone marrow

## Humoral (Antibody-Mediated) Immune Response

**23.** The basic structure of an antibody molecule is diagrammed in Figure 12–5. Select different colors, and color in the coding circles below and the corresponding areas on the diagram.

◯ heavy chains        ◯ light chains

Add labels to the diagram to correctly identify the type of bonds holding the polypeptide chains together. Also label the constant (C) and variable (V) regions of the antibody, and add "polka dots" to the variable portions.

**Figure 12–5**

1. Which portion of the antibody—V or C—is its antigen-binding site?

_____

2. Which portion acts to determine antibody class and specific function?

_____

**24.** Match the antibody classes in Column B to their descriptions in Column A. Place the correct term(s) or letter response(s) in the answer blanks.

**Column A**

**Column B**

_____ 1. Bound to the surface of a B cell          A. IgA

_____ 2. Crosses the placenta          B. IgD

_____ 3. The first antibody released during the          C. IgE
primary response

_____ 4. Fixes complement (two classes)          D. IgG

_____ 5. Is a pentamer          E. IgM

_____ 6. The most abundant antibody found in
blood plasma and the chief antibody
released during secondary responses

_____ 7. Binds to the surface of mast cells and
mediates an allergic response

_____ 8. Predominant antibody found in mucus,
saliva, and tears

**25.** Complete the following descriptions of antibody function by writing the missing terms in the answer blanks.

_____ 1.  Antibodies can inactivate antigens in various ways, depending
on the nature of the __(1)__. __(2)__ is the chief ammunition
_____ 2.  used against cellular antigens such as bacteria and mismatched
red blood cells. The binding of antibodies to sites on bacterial
_____ 3.  exotoxins or viruses that can cause cell injury is called __(3)__.
The cross linking of cellular antigens into large lattices by
_____ 4.  antibodies is called __(4)__; Ig __(5)__, with its 10 antigen binding sites, is particularly efficient in this mechanism. When
_____ 5.  molecules are cross-linked into lattices by antibodies, the
mechanism is more properly called __(6)__. In virtually all
_____ 6.  these cases, the protective mechanism mounted by the antibodies serves to disarm and/or immobilize the antigens until
_____ 7.  they can be disposed of by __(7)__.

26. Determine whether each of the following situations provides, or is an example of, active or passive immunity. If passive, write *P* in the blank; if active, write *A* in the blank.

_____ 1. An individual receives Sabin polio vaccine

_____ 2. Antibodies migrate through a pregnant woman's placenta into the vascular system of her fetus

_____ 3. A student nurse receives an injection of gamma globulin (containing antibodies to the hepatitis virus) after she has been exposed to viral hepatitis

_____ 4. "Borrowed" immunity

_____ 5. Immunologic memory is provided

_____ 6. An individual suffers through chicken pox

27. There are several important differences between primary and secondary immune response(s). If the following statements best describe a primary response, write *P* in the blank; if a secondary response, write *S* in the blank.

_____ 1. The initial response to an antigen; gearing-up stage

_____ 2. A lag period of several days occurs before antibodies specific to the antigen appear in the bloodstream

_____ 3. Antibody levels increase rapidly and remain high for an extended period

_____ 4. Immunologic memory is established

_____ 5. The second, third, and subsequent responses to the same antigen

## Cellular (Cell-Mediated) Immune Response

28. Several populations of T cells exist. Match the terms in Column B to the descriptions in Column A. Place the correct term or letter response in the answer blanks.

**Column A**

_____ 1. Binds with B cells and releases chemicals that activate B cells, T cells, and macrophages

_____ 2. Activated by recognizing both its antigen and a self-protein presented on the surface of a macrophage

_____ 3. Turns off the immune response when the "enemy" has been routed

_____ 4. Directly attacks and lyses cellular pathogens

**Column B**

A. Helper T cell

B. Killer T cell

C. Suppressor T cell

**29.** Using the key choices, select the terms that correspond to the descriptions of substances or events by inserting the appropriate term or letter in the answer blanks.

*Key Choices*

A. Anaphylactic shock     D. Complement          G. Interferon

B. Antibodies             E. Cytokines           H. Lymphokine

C. Chemotaxis factors     F. Inflammation        I. Monokines

_____ 1. A protein released by macrophages and activated T cells that helps to protect other body cells from viral multiplication

_____ 2. Any types of molecules that attract neutrophils and other protective cells into a region where an immune response is ongoing

_____ 3. Proteins released by plasma cells that mark antigens for destruction by phagocytes or complement

_____ 4. A consequence of the release of histamine and of complement activation

_____ 5. C and G are examples of this class of molecules

_____ 6. A group of plasma proteins that amplifies the immune response by causing lysis of cellular pathogens once it has been "fixed" to their surface

_____ 7. Class of chemicals released by macrophages

**30.** Organ transplants are often unsuccessful because self-proteins vary in different individuals. However, chances of success increase if certain important procedures are followed. The following questions refer to this important area of clinical medicine.

1. Assuming that autografts and isografts are not possible, what is the next most successful

   graft type and what is its source? _____

2. What two cell types are important in rejection phenomena?

   _____

3. Why are immunosuppressive drugs (or therapy) provided after transplant surgery and what is the major shortcoming of this therapy?

   _____

   _____

   _____

**31.** Figure 12–6 is a flowchart of the immune response that tests your understanding of the interrelationships of that process. Several terms have been omitted from this schematic. First, complete the figure by inserting appropriate terms from the key choices below. (Note that oval blanks indicate that the required term identifies a cell type and rectangular blanks represent the names of chemical molecules. Also note that solid lines represent stimulatory or enhancing effects, whereas broken lines indicate inhibition.) Then color the coding circles and the corresponding ovals, indicating the cell types identified.

*Key Choices*

Cell types:

◯ B cell

◯ Helper T cell

◯ Killer T cell

◯ Macrophage

◯ Memory B cell

◯ Memory T cell

◯ Neutrophils

◯ Plasma cell

◯ Suppressor T cell

Molecules:

Antibodies

Chemotaxis factors

Complement

Interferon

Lymphokines

Monokines

Perforin

Suppressor factors

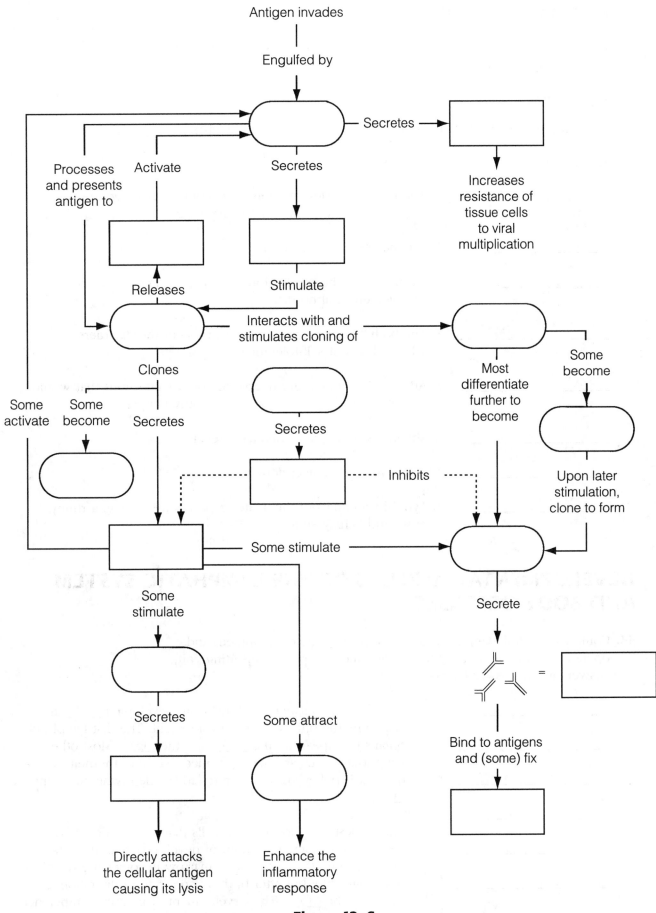

Antigen invades

Engulfed by

Secretes

Increases
resistance of
tissue cells
to viral
multiplication

Processes
and presents
antigen to

Activate

Secretes

Releases

Stimulate

Interacts with and
stimulates cloning of

Most
differentiate
further to
become

Some
become

Clones

Some
activate

Some
become

Secretes

Secretes

Inhibits

Upon later
stimulation,
clone to form

Some stimulate

Secrete

Some
stimulate

Some attract

Bind to antigens
and (some) fix

Secretes

Directly attacks
the cellular antigen
causing its lysis

Enhance the
inflammatory
response

**Figure 12-6**

# Disorders of Immunity

**32.** Using the key choices, identify the type of immunity disorder described. Insert the appropriate term or letter in the answer blank.

*Key Choices*

A. Allergy          B. Autoimmune disease          C. Immunodeficiency

_____ 1. AIDS and SCID

_____ 2. The immune system mounts an extraordinarily vigorous response to an otherwise harmless antigen

_____ 3. A hypersensitivity reaction

_____ 4. Occurs when the production or activity of immune cells or complement is abnormal

_____ 5. The body's own immune system produces the disorder; a breakdown of self-tolerance

_____ 6. Affected individuals are unable to combat infections that would present no problem for normally healthy people

_____ 7. Multiple sclerosis and rheumatic fever

_____ 8. Hay fever and contact dermatitis

_____ 9. Typical symptoms of the acute response are tearing, a runny nose, and itching skin

# DEVELOPMENTAL ASPECTS OF THE LYMPHATIC SYSTEM AND BODY DEFENSES

**33.** Complete the following statements concerning the development and operation of the immune system during the life span by inserting your answers in the answer blanks.

_____ 1.

_____ 2.

_____ 3.

_____ 4.

_____ 5.

_____ 6.

_____ 7.

Lymphatic vessels which "bud" from developing __(1)__ are visible by the fifth week of development. The first lymphoid organs to appear are the __(2)__ and the __(3)__. Most other lymphoid organs are poorly formed before birth; their development is believed to be controlled by the thymic hormone __(4)__.

The earliest lymphocyte stem cells that can be identified appear during the first month of development in the fetal __(5)__. Shortly thereafter, bone marrow becomes the lymphocyte origin site; but after birth, lymphocyte proliferation occurs in the __(6)__. The development of immunocompetence has usually been accomplished by __(7)__.

_____ 8.

_____ 9.

_____ 10.

_____ 11.

_____ 12.

During old age, the effectiveness of the immune system __(8)__, and elders are more at risk for __(9)__, __(10)__, and __(11)__. Part of the declining defenses may reflect the fact that __(12)__ antibodies are unable to get to the mucosal surfaces where they carry out their normal protective role.

## INCREDIBLE JOURNEY

### A Visualization Exercise for the Immune System

*Something quite enormous and looking much like an octopus is nearly blocking the narrow tunnel just ahead.*

**34.** Where necessary, complete statements by inserting the missing word(s) in the answer blanks.

_____ 1.

_____ 2.

_____ 3.

For this journey, you are equipped with scuba gear before you are miniaturized and injected into one of your host's lymphatic vessels. He has been suffering with a red, raw "strep throat" and has swollen cervical lymph nodes. Your assignment is to travel into a cervical lymph node and observe the activities going on there that reveal that your host's immune system is doing its best to combat the infection.

On injection, you enter the lymph with a "WHOOSH" and then bob gently in the warm yellow fluid. As you travel along, you see what seem to be thousands of spherical bacteria and a few large globular __(1)__ molecules that, no doubt, have been picked up by the tiny lymphatic capillaries. Shortly thereafter, a large dark mass, shaped like a kidney bean, looms just ahead. This has to be a __(2)__, you conclude, and you dig in your wet suit pocket to find the waterproof pen and recording tablet.

As you enter the gloomy mass, the lymphatic stream becomes shallow and begins to flow sluggishly. So that you can explore this little organ fully, you haul yourself to your feet and begin to wade through the slowly moving stream. On each bank you see a huge ball of cells that have large nuclei and such a scant amount of cytoplasm that you can barely make it out. You write, "Sighted the spherical germinal centers composed of __(3)__." As you again study one of the cell masses, you spot one cell that looks quite different and reminds you of a nest of angry hornets, because it is furiously spewing out what seems to be a horde of tiny Y-shaped "bees." "Ah ha,"

_____  4.

_____  5.

_____  6.

_____  7.

_____  8.

_____  9.

_____  10.

_____  11.

you think, "another valuable piece of information." You record, "Spotted a __(4)__ making and releasing __(5)__."

That done, you turn your attention to scanning the rest of the landscape. Suddenly you let out an involuntary yelp. Something quite enormous and looking much like an octopus is nearly blocking the narrow tunnel just ahead. Your mind whirls as it tries to figure out the nature of this cellular "beast" that appears to be guarding the channel. Then it hits you—this has to be a __(6)__ on the alert for foreign invaders (more properly called __(7)__ ), which it "eats" when it catches them. The giant cell roars, "Halt, stranger, and be recognized," and you dig frantically in your pocket for your identification pass. As you drift toward the huge cell, you hold the pass in front of you, hands trembling because you know this cell could liquefy you quick as the blink of an eye. Again the cell bellows at you, "Is this some kind of a security check? I'm on the job, as you can see!" Frantically you shake your head "NO," and the cell lifts one long tentacle and allows you to pass. As you squeeze by, the cell says, "Being inside, I've never seen my body's outside. I must say, humans are a rather strange-looking lot!" Still shaking, you decide that you are in no mood for a chat and hurry along to put some distance between yourself and this guard cell.

Immediately ahead are what appear to be hundreds of the same type of cell sitting on every ledge and in every nook and cranny. Some are busily snagging and engulfing unfortunate strep bacteria that float too close. The slurping sound is nearly deafening. Then something grabs your attention: The surface of one of these cells is becoming dotted with some of the same donut-shaped chemicals that you see on the strep bacteria membranes; and a round cell, similar, but not identical, to those you earlier saw in the germinal centers, is starting to bind to one of these "doorknobs." You smile smugly, because you know you have properly identified the octopus-like cells. You then record your observations as follows: "Cells like the giant cell just identified act as __(8)__. I have just observed one in this role during its interaction with a helper __(9)__ cell."

You decide to linger a bit to see if the round cell becomes activated. You lean against the tunnel walls and watch quietly, but your wait is brief. Within minutes, the cell that was binding to the octopus-like cell begins to divide, and then its daughter cells divide again and again at a head-spinning pace. You write, "I have just witnessed the formation of a __(10)__ of like cells." Most of the daughter cells enter the lymph stream, but a few of them settle back and seem to go into a light sleep. You decide that the "napping cells" don't have any role to play in helping get rid of your host's present strep infection, but instead will provide for __(11)__ and become active at a later date.

You glance at your watch and wince as you realize that it is already 5 minutes past the time for your retrieval. You have already concluded that this is a dangerous place for those who don't "belong" and are far from sure about how long your pass is good, so you swim hurriedly from the organ into the lymphatic stream to reach your pickup spot.

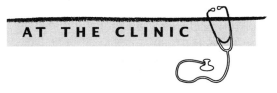

**AT THE CLINIC**

**35.** A young man is rushed to the emergency room after fainting. His blood pressure is alarmingly low, and his companion reports the man collapsed shortly after being stung by a wasp. What has caused his hypotension? What treatment will be given immediately?

**36.** Patty Hourihan is a strict environmentalist and a new mother. Although she is very much against using disposable diapers, she is frustrated by the fact that her infant breaks out in a diaper rash when she uses cloth diapers. Since new cloth diapers do not cause the rash, but washed ones do, what do you think the problem is?

**37.** James, a 36-year-old engineer, appeared at the clinic in an extremely debilitated condition. He had purple-brown lesions on his skin and a persistent cough. A physical examination revealed swollen lymph nodes. Laboratory tests revealed a low lymphocyte count. Information taken during the personal history revealed that James was homosexual. The skin lesions proved to be evidence of Kaposi's sarcoma. What is James's problem?

**38.** About 6 months after an automobile accident in which her neck was severely lacerated, a young woman comes to the clinic for a routine checkup. Visual examination shows a slight swelling just inferior to her larynx; her skin is dry and her face is puffy. When questioned, the woman reports that she fatigues easily, has been gaining weight, and her hair is falling out. What do you think is wrong?

**39.** Young Joe Chang went sledding and the runner of a sled hit him in the left side and ruptured his spleen. Joe almost died because he did not get to the hospital fast enough. Upon arrival, a splenectomy was performed. What, would you guess, is the immediate danger of spleen rupture? Will Joe require a transplant for spleen replacement?

**40.** Use of birth control pills decreases the acidity of the vaginal tract. Why might this increase the incidence of vaginal infection (vaginitis)?

**41.** After surgery to remove lymphatic vessels associated with the removal of a melanoma, what condition can be expected relative to lymph drainage? Is this a permanent problem? Why or why not?

CHAPTER

13

# The Respiratory System

Body cells require an abundant and continuous supply of oxygen to carry out their activities. As cells use oxygen, they release carbon dioxide, a waste product that must be eliminated from the body. The circulatory and respiratory systems are intimately involved in obtaining and delivering oxygen to body cells and in eliminating carbon dioxide from the body. The respiratory system is responsible for gas exchange between the pulmonary blood and the external environment (that is, external respiration). The respiratory system also plays an important role in maintaining the acid-base balance of the blood.

Questions and activities in this chapter consider both the anatomy and physiology of the respiratory system structures.

## FUNCTIONAL ANATOMY

1. The following questions refer to the primary bronchi. In the spaces provided, insert the letter *R* to indicate the right primary bronchus and the letter *L* to indicate the left primary bronchus.

   1. Which of the primary bronchi is larger in diameter? _____

   2. Which of the primary bronchi is more horizontal? _____

   3. Which of the primary bronchi is the most common site for lodging of a foreign object

      that has entered the respiratory passageways? _____

2. Complete the following statements by inserting your answers in the answer blanks.

_____ 1.

_____ 2.

_____ 3.

_____ 4.

_____ 5.

_____ 6.

_____ 7.

_____ 8.

_____ 9.

_____ 10.

_____ 11.

_____ 12.

_____ 13.

_____ 14.

_____ 15.

_____ 16.

Air enters the nasal cavity of the respiratory system through the  (1) . The nasal cavity is divided by the midline  (2) . The nasal cavity mucosa has several functions. Its major functions are to  (3) ,  (4) , and  (5)  the incoming air. Mucous membrane–lined cavities called  (6)  are found in several bones surrounding the nasal cavities. They make the skull less heavy and probably act as resonance chambers for  (7) . The passageway common to the digestive and respiratory systems, the  (8) , is often referred to as the throat; it connects the nasal cavity with the  (9)  below. Clusters of lymphatic tissue,  (10) , are part of the defensive system of the body. Reinforcement of the trachea with  (11)  rings prevents its collapse during  (12)  changes that occur during breathing. The fact that the rings are incomplete posteriorly allows a food bolus to bulge  (13)  during its transport to the stomach. The larynx or voice box is built from many cartilages, but the largest is the  (14)  cartilage. Within the larynx are the  (15) , which vibrate with exhaled air and allow an individual to  (16) .

3. Circle the term that does not belong in each of the following groupings.

1. Sphenoidal      Maxillary      Mandibular      Ethmoid      Frontal

2. Nasal cavity      Trachea      Alveolus      Larynx      Bronchus

3. Apex      Base      Hilus      Larynx      Pleura

4. Sinusitis      Peritonitis      Pleurisy      Tonsillitis      Laryngitis

5. Laryngopharynx      Oropharynx      Transports air and food      Nasopharynx

6. Alveoli      Respiratory zone      Alveolar sac      Primary bronchus

**4.** Figure 13–1 is a sagittal view of the upper respiratory structures. First, correctly identify all structures provided with leader lines on the figure. Then select different colors for the structures listed below and use them to color in the coding circles and the corresponding structures on the figure.

◯ Nasal cavity          ◯ Larynx

◯ Pharynx              ◯ Paranasal sinuses

◯ Trachea

**Figure 13–1**

**5.** Using the key choices, select the terms identified in the following descriptions by inserting the appropriate term or letter in the answer blanks.

*Key Choices*

A. Alveoli        D. Epiglottis        G. Palate            J. Primary bronchi

B. Bronchioles    E. Esophagus         H. Parietal pleura   K. Trachea

C. Conchae        F. Glottis           I. Phrenic nerve     L. Visceral pleura

_____ 1. Smallest conducting respiratory passageways

_____ 2. Separates the oral and nasal cavities

_____ 3. Major nerve, stimulating the diaphragm

_____ 4. Food passageway posterior to the trachea

_____ 5. Closes off the larynx during swallowing

_____ 6. Windpipe

_____ 7. Actual site of gas exchanges

_____ 8. Pleural layer covering the thorax walls

_____ 9. Pleural layer covering the lungs

_____ 10. Lumen of larynx

_____ 11. Fleshy lobes in the nasal cavity which increase its surface area

**6.** Complete the following paragraph concerning the alveolar cells and their roles by writing the missing terms in the answer blanks.

_____ 1.

_____ 2.

_____ 3.

_____ 4.

With the exception of the stroma of the lungs, which is __(1)__ tissue, the lungs are mostly air spaces, of which the alveoli comprise the greatest part. The bulk of the alveolar walls are made up of squamous epithelial cells, which are well suited for their __(2)__ function. Much less numerous cuboidal cells produce a fluid that coats the air-exposed surface of the alveolus and contains a lipid-based molecule called __(3)__ that functions to __(4)__ of the alveolar fluid.

**7.** Figure 13–2 is a diagram of the larynx and associated structures. On the figure, identify each of the structures listed below. Select a different color for each and use it to color in the coding circles and the corresponding structures on the figure. Then answer the questions following the diagram.

◯ Hyoid bone          ◯ Tracheal cartilages          ◯ Circoid cartilage

◯ Thyroid cartilage          ◯ Epiglottis

Ligaments

**Figure 13–2**

1. What are three functions of the larynx? _____

_____

_____

2. What type of cartilage forms the epiglottis? _____

3. What type of cartilage forms the other eight laryngeal cartilages? _____

4. Explain this difference. _____

_____

_____

5. What is the common name for the thyroid cartilage?_____

**8.** Figure 13–3 illustrates the gross anatomy of the lower respiratory system. Intact structures are shown on the left; respiratory passages are shown on the right. Select a different color for each of the structures listed below and use it to color in the coding circles and the corresponding structures on the figure. Then complete the figure by labeling the areas/structures that are provided with leader lines on the figure. Be sure to include the following: pleural space, mediastinum, apex of right lung, diaphragm, clavicle, and the base of the right lung.

◯ Trachea           ◯ Primary bronchi      ◯ Visceral pleura

◯ Larynx            ◯ Secondary bronchi    ◯ Parietal pleura

◯ Intact lung

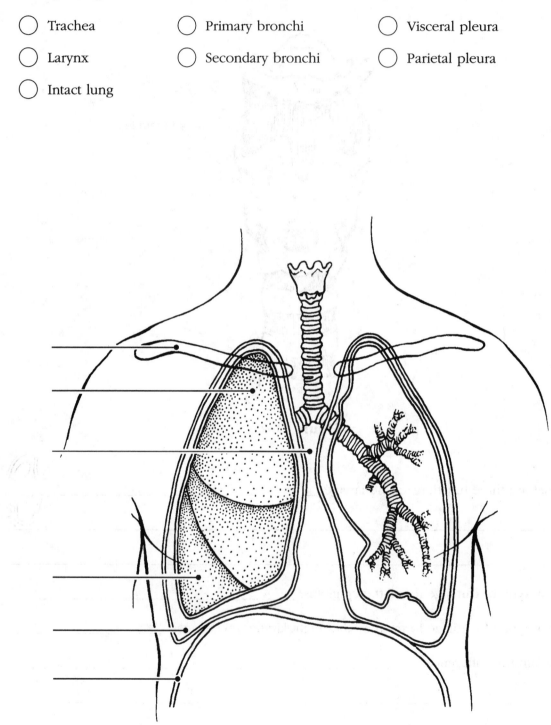

**Figure 13–3**

**9.** Figure 13–4 illustrates the microscopic structure of the respiratory unit of lung tissue. The external anatomy is shown in Figure 13–4A. Color the intact alveoli yellow, the pulmonary capillaries red, and the respiratory bronchioles green.

A cross section through an alveolus is shown on Figure 13–4B and a blow-up of the respiratory membrane is shown in Figure 13–4C. On these illustrations color the alveolar epithelium yellow, the capillary endothelium pink, and the red blood cells in the capillary red. Also, label the alveolar chamber and color it pale blue. Finally, in Figure 13–4C label the region of the fused basement membranes; add the symbols for oxygen gas ($O_2$) and carbon dioxide gas ($CO_2$) in the sites where they would be in higher concentration and arrows correctly showing their direction of movement through the respiratory membrane.

**Figure 13–4**

A

B

C

Respiratory membrane

# RESPIRATORY PHYSIOLOGY

**10.** Using the key choices, select the terms identified in the following descriptions by inserting the appropriate term or letter in the answer blanks.

*Key Choices*

A. Atmospheric pressure        B. Intrapulmonary pressure        C. Intrapleural pressure

_____ 1. In healthy lungs, it is always lower than atmospheric pressure (that is, it is negative pressure)

_____ 2. Pressure of air outside the body

_____ 3. As it decreases, air flows into the passageways of the lungs

_____ 4. As it increases over atmospheric pressure, air flows out of the lungs

_____ 5. If this pressure becomes equal to the atmospheric pressure, the lungs collapse

_____ 6. Rises well over atmospheric pressure during a forceful cough

**11.** Many changes occur within the lungs as the diaphragm (and external intercostal muscles) contract and then relax. These changes lead to the flow of air into and out of the lungs. The activity of the diaphragm is given in the left column of the following table. Several changes in condition are listed in the column heads to the right. Complete the table by checking (✓) the appropriate column to correctly identify the change that would be occurring relative to the diaphragm's activity in each case.

| **Activity of diaphragm** | **Changes in** | | | | | | | |
|---|---|---|---|---|---|---|---|---|
| | **Internal volume of thorax** | | **Internal pressure in thorax** | | **Size of lungs** | | **Direction of air flow** | |
| (↑ = increased) (↓ = decreased) | ↑ | ↓ | ↑ | ↓ | ↑ | ↓ | Into lung | Out of lung |
| Contracted, moves downward | | | | | | | | |
| Relaxed, moves superiorly | | | | | | | | |

**12.** Use the key choices to respond to the following descriptions. Insert the correct term or letter in the answer blanks.

*Key Choices*

A. External respiration      C. Inspiration      E. Ventilation (breathing)

B. Expiration      D. Internal respiration

_____ 1. Period of breathing when air enters the lungs

_____ 2. Exchange of gases between the systemic capillary blood and body cells

_____ 3. Alternate flushing of air into and out of the lungs

_____ 4. Exchange of gases between alveolar air and pulmonary capillary blood

**13.** Although normal quiet expiration is largely passive due to lung recoil, when expiration must be more forceful (or the lungs are diseased), muscles that increase the abdominal pressure or depress the rib cage are enlisted.

1. Provide two examples of muscles that cause abdominal pressure to rise.

_____ and _____

2. Provide two examples of muscles that depress the rib cage.

_____ and _____

**14.** Four nonrespiratory movements are described here. Identify each by inserting your answers in the spaces provided.

1. Sudden inspiration, resulting from spasms of the diaphragm. _____

2. A deep breath is taken, the glottis is closed, and air is forced out of the lungs against

the glottis; clears the lower respiratory passageways. _____

3. As just described, but clears the upper respiratory passageways. _____

4. Increases ventilation of the lungs; may be initiated by a need to increase oxygen levels

in the blood. _____

**15.** The following section concerns respiratory volume measurements. Using key choices, select the terms identified in the following descriptions by inserting the appropriate term or letter in the answer blanks.

*Key Choices*

A. Dead space volume     C. Inspiratory reserve volume (IRV)     E. Tidal volume (TV)

B. Expiratory reserve     D. Residual volume (RV)     F. Vital capacity (VC)
    volume (ERV)

_____ 1. Respiratory volume inhaled or exhaled during normal breathing

_____ 2. Air in respiratory passages that does not contribute to gas exchange

_____ 3. Total amount of exchangeable air

_____ 4. Gas volume that allows gas exchange to go on continuously

_____ 5. Amount of air that can still be exhaled (forcibly) after a normal exhalation

**16.** Figure 13–5 is a diagram showing respiratory volumes. Complete the figure by making the following additions.

1. Bracket the volume representing the vital capacity and color the area yellow; label it VC.

2. Add green stripes to the area representing the inspiratory reserve volume and label it IRV.

3. Add red stripes to the area representing the expiratory reserve volume and label it ERV.

4. Identify and label the respiratory volume, which is *now just yellow*. Color the residual volume (RV) blue and label it appropriately on the figure.

5. Bracket and label the inspiratory capacity (IC).

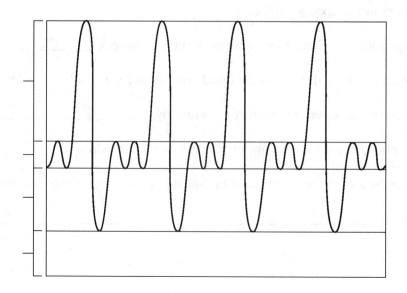

**Figure 13–5**

**17.** Use the key choices to correctly complete the following statements, which refer to gas exchanges in the body. Insert the correct letter response in the answer blanks.

***Key Choices***

A. Active transport

B. Air of alveoli to capillary blood

C. Carbon dioxide-poor and oxygen-rich

D. Capillary blood to alveolar air

E. Capillary blood to tissue cells

F. Diffusion

G. Higher concentration

H. Lower concentration

I. Oxygen-poor and carbon dioxide-rich

J. Tissue cells to capillary blood

_____ 1.

_____ 2.

_____ 3.

_____ 4.

_____ 5.

_____ 6.  _____ 7.  _____ 8.  _____ 9.

All gas exchanges are made by  (1) . When substances pass in this manner, they move from areas of their  (2)  to areas of their  (3) . Thus oxygen continually passes from the  (4)  and then from the  (5) . Conversely, carbon dioxide moves from the  (6)  and from  (7) . From there it passes out of the body during expiration. As a result of such exchanges, arterial blood tends to be  (8)  while venous blood is  (9) .

**18.** Complete the following statements by inserting your answers in the answer blanks.

_____ 1.

_____ 2.

_____ 3.

_____ 4.

Most oxygen is transported bound to  (1)  inside the red blood cells. Conversely, *most* carbon dioxide is carried in the form of  (2)  in the  (3) . Carbon monoxide poisoning is lethal because carbon monoxide competes with  (4)  for binding sites.

**19.** Circle the term that does not belong in each of the following groupings.

1. ↑ Respiratory rate     ↓ In blood $CO_2$     Alkalosis     Acidosis

2. Acidosis     ↑ Carbonic acid     ↓ pH     ↑ pH

3. Acidosis     Hyperventilation     Hypoventilation     $CO_2$ buildup

4. Apnea     Cyanosis     ↑ Oxygen     ↓ Oxygen

5. ↑ Respiratory rate     ↑ Exercise     Anger     ↑ $CO_2$ in blood

6. High altitude     ↓ $PO_2$     ↑ $PCO_2$     ↓ Atmospheric pressure

**20.** There are several levels of breathing control. Match the structures given in Column B to the appropriate descriptions provided in Column A. Place the correct term or letter response in the answer blanks provided.

### Column A                                                    Column B

_____ 1. Smooth out the basic rhythm of breath-        A. Chemoreceptors
                      ing set by the medulla                          in the aortic and
                                                                      carotid bodies

_____ 2. Respiratory control center in the medulla     B. Intercostal

_____ 3. Respond to overinflation of the lungs         C. Inspiratory center

_____ 4. Respond to decreased oxygen levels            D. Phrenic
                      in the blood
                                                                   E. Pons centers
_____ , _____ 5. Nerves that carry activating impulses
                      to the muscles of inspiration                F. Stretch receptors
                                                                      in the lungs

## RESPIRATORY DISORDERS

**21.** Match the terms in Column B with the pathologic conditions described in Column A.

### Column A                                                    Column B

_____ 1. Lack or cessation of breathing               A. Apnea

_____ 2. Normal breathing in terms of rate and        B. Asthma
                      depth
                                                                  C. Chronic bronchitis
_____ 3. Labored breathing, or "air hunger"
                                                                  D. Dyspnea
_____ 4. Chronic oxygen deficiency
                                                                  E. Emphysema
_____ 5. Condition characterized by fibrosis of
                      the lungs and an increase in size of the    F. Eupnea
                      alveolar chambers
                                                                  G. Hypoxia
_____ 6. Condition characterized by increased
                      mucus production, which clogs respira-      H. Lung cancer
                      tory passageways and promotes coughing
                                                                  I. Tuberculosis
_____ 7. Respiratory passageways narrowed
                      by bronchiolar spasms

_____ 8. Together called COPD

_____ 9. Incidence strongly associated with cigarette smoking; outlook is poor

_____ 10. Infection spread by airborne bacteria; a recent alarming increase
                      in drug users and AIDs victims

# DEVELOPMENTAL ASPECTS OF THE RESPIRATORY SYSTEM

22. Mrs. Jones gave birth prematurely to her first child. At birth, the baby weighed 2 lb 8 oz. Within a few hours, the baby had developed severe dyspnea and was becoming cyanotic. Therapy with a positive pressure ventilator was prescribed. Answer the following questions related to the situation just described. Place your responses in the answer blanks.

   1. The infant's condition is referred to as _____

      _____

   2. It occurs because of a relative lack of _____

   3. The function of the deficient substance is to _____

      _____

   4. Explain what the positive pressure apparatus accomplishes. _____

      _____

23. Complete the following statements by inserting your answers in the answer blanks.

    _____ 1.

    _____ 2.

    _____ 3.

    _____ 4.

    _____ 5.

    _____ 6.

    _____ 7.

    _____ 8.

    The respiratory rate of a newborn baby is approximately __(1)__ respirations per minute. In a healthy adult, the respiratory range is __(2)__ respirations per minute. Most problems that interfere with the operation of the respiratory system fall into one of the following categories: infections such as pneumonia, obstructive conditions such as __(3)__ and __(4)__, and/or conditions that destroy lung tissue, such as __(5)__. With age, the lungs lose their __(6)__, and the __(7)__ of the lungs decreases. Protective mechanisms also become less efficient, causing elderly individuals to be more susceptible to __(8)__.

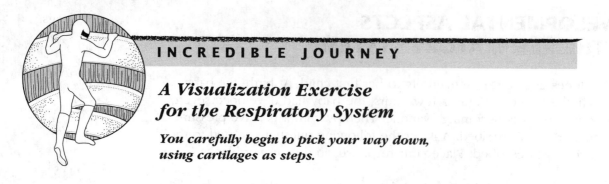

*A Visualization Exercise for the Respiratory System*

**You carefully begin to pick your way down, using cartilages as steps.**

**24.** Where necessary, complete statements by inserting the missing word(s) in the answer blanks.

_____ 1.

_____ 2.

_____ 3.

_____ 4.

_____ 5.

_____ 6.

_____ 7.

_____ 8.

_____ 9.

_____ 10.

_____ 11.

_____ 12.

_____ 13.

Your journey through the respiratory system is to be on foot. To begin, you simply will walk into your host's external nares. You are miniaturized, and your host is sedated lightly to prevent sneezing during your initial observations in the nasal cavity and subsequent descent.

You begin your exploration of the nasal cavity in the right nostril. One of the first things you notice is that the chamber is very warm and humid. High above, you see three large, round lobes, the _(1)_, which provide a large mucosal surface area for warming and moistening the entering air. As you walk toward the rear of this chamber, you see a large lumpy mass of lymphatic tissue, the _(2)_ in the _(3)_, or first portion of the pharynx. As you peer down the pharynx, you realize that it will be next to impossible to maintain your footing during the next part of your journey. It is nearly straight down, and the _(4)_ secretions are like grease. You sit down and dig your heels in to get started. After a quick slide, you land abruptly on one of a pair of flat, sheetlike structures that begin to vibrate rapidly, bouncing you up and down helplessly. You are also conscious of a rhythmic hum during this jostling, and you realize that you have landed on a _(5)_. You pick yourself up and look over the superior edge of the _(6)_, down into the seemingly endless esophagus behind. You chastize yourself for not remembering that the _(7)_ and respiratory pathways separate at this point. Hanging directly over your head is the leaflike _(8)_ cartilage.

Normally, you would not have been able to get this far because it would have closed off this portion of the respiratory tract. With your host sedated, however, that protective reflex does not work.

You carefully begin to pick your way down, using the cartilages as steps. When you reach the next respiratory organ, the _(9)_, your descent becomes much easier, because the structure's C-shaped cartilages form a ladderlike supporting structure. As you climb down the cartilages, your face is stroked rhythmically by soft cellular extensions, or _(10)_. You remember that their function is to move mucus laden with bacteria or dust and other debris toward the _(11)_.

You finally reach a point where the descending passageway splits into two _(12)_, and since you want to control your progress (rather than slide downward), you choose the more horizontal _(13)_

_____ 14.

_____ 15.

_____ 16.

_____ 17.

_____ 18.

_____ 19.

_____ 20.

branch. If you remain in the superior portion of the lungs, your return trip will be less difficult because the passageways will be more horizontal than steeply vertical. The passageways get smaller and smaller, slowing your progress. As you are squeezing into one of the smallest of the respiratory passageways, a (14), you see a bright spherical chamber ahead. You scramble into this (15), pick yourself up, and survey the area. Scattered here and there are lumps of a substance that looks suspiciously like coal, reminding you that your host is a smoker. As you stand there, a soft rustling wind seems to flow in and out of the chamber. You press your face against the transparent chamber wall and see disc-like cells, (16), passing by in the capillaries on the other side. As you watch, they change from a somewhat bluish color to a bright (17) color as they pick up (18) and unload (19).

You record your observations and then contact headquarters to let them know you are ready to begin your ascent. You begin your return trek, slipping and sliding as you travel. By the time you reach the inferior edge of the trachea, you are ready for a short break. As you rest on the mucosa, you begin to notice that the air is becoming close and very heavy. You pick yourself up quickly and begin to scramble up the trachea. Suddenly and without warning, you are hit by a huge wad of mucus and catapulted upward and out onto your host's freshly pressed handkerchief! Your host has assisted your exit with a (20).

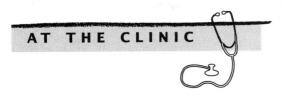

## AT THE CLINIC

**25.** After a long bout of bronchitis, Ms. Dupee complains of a stabbing pain in her side with each breath. What is her probable condition?

**26.** The Kozloski family is taking a long auto trip. Michael, who has been riding in the back of the station wagon, complains of a throbbing headache. A little later, he seems confused and his face is flushed. What is your diagnosis of Michael's problem?

**27.** A new mother checks on her sleeping infant, only to find that it has stopped breathing and is turning blue. The mother quickly picks up the baby and pats its back until it starts to breathe. What tragedy has been averted?

**28.** Joanne Willis, a long-time smoker, is complaining that she has developed a persistent cough. What is your first guess as to her condition? What has happened to her bronchial cilia?

**29.** Barbara is rushed to the emergency room after an auto accident. The 8th through 10th ribs on her left side have been fractured and have punctured the lung. What term is used to indicate lung collapse? Will both lungs collapse? Why or why not?

**30.** A young boy is diagnosed with cystic fibrosis. What effect will this have on his respiratory system?

**31.** Mr. and Ms. Rao took their sick 5-year-old daughter to the doctor. The girl was breathing entirely through her mouth, her voice sounded odd and whiny, and a puslike fluid was dripping from her nose. Which one of the sets of tonsils was most likely infected in this child?

**32.** Assume that you are a second-year nursing student. As your assignment you are asked to explain how each of the following might interfere with a patient's gas exchange.

1. Iron deficiency that causes a decrease in the number of red blood cells.

2. Cystic fibrosis in which the surfaces of the alveoli become coated with thick, sticky mucus.

3. The patient is a heavy smoker.

# 14 The Digestive System and Body Metabolism

The digestive system processes food so that it can be absorbed and used by the body's cells. The digestive organs are responsible for food ingestion, digestion, absorption, and elimination of undigested remains from the body. In one sense, the digestive tract can be viewed as a disassembly line in which food is carried from one stage of its breakdown process to the next by muscular activity, and its nutrients are made available en route to the cells of the body. In addition, the digestive system provides for one of life's greatest pleasures—eating.

The anatomy of both alimentary canal and accessory digestive organs, mechanical and enzymatic breakdown, and absorption mechanisms are covered in this chapter. An introduction to nutrition and some important understandings about cellular metabolism (utilization of foodstuffs by body cells) are also considered in this chapter review.

## ANATOMY OF THE DIGESTIVE SYSTEM

1. Complete the following statements by inserting your answers in the answer blanks.

   _____ 1.

   _____ 2.

   _____ 3.

   _____ 4.

   _____ 5.

   _____ 6.

   _____ 7.

   The digestive system is responsible for many body processes. Its functions begin when food is taken into the mouth, or __(1)__. The process called __(2)__ occurs as food is broken down both chemically and mechanically. For the broken-down foods to be made available to the body cells, they must be absorbed through the digestive system walls into the __(3)__. Undigestible food remains are removed, or __(4)__, from the body in __(5)__. The organs forming a continuous tube from the mouth to the anus are collectively called the __(6)__. Organs located outside the digestive tract proper, which secrete their products into the digestive tract, are referred to as __(7)__ digestive system organs.

2. Figure 14–1 is a frontal view of the digestive system. First, correctly identify all structures provided with leader lines. Then select different colors for the following organs and color the coding circles and the corresponding structures of the figure.

◯ Esophagus       ◯ Pancreas          ◯ Tongue

◯ Liver           ◯ Salivary glands   ◯ Uvula

◯ Large intestine ◯ Small intestine

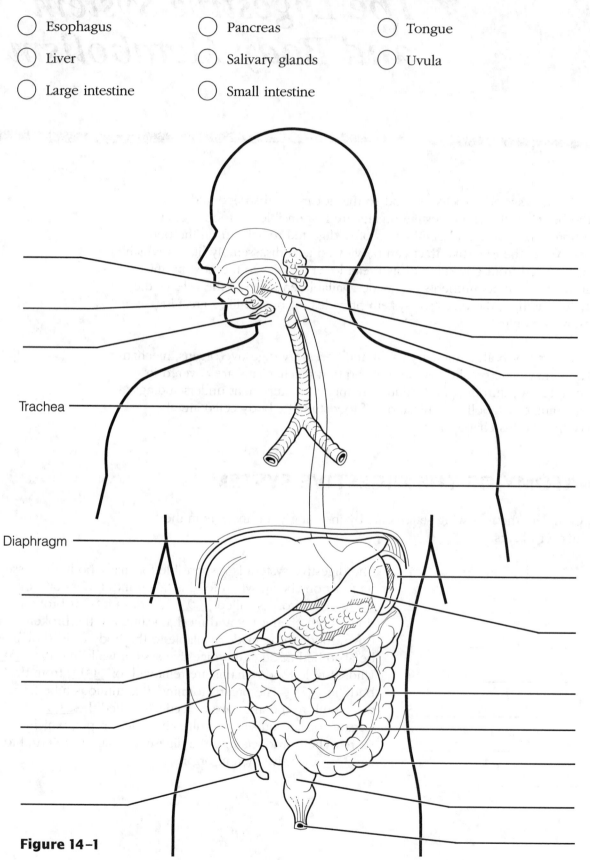

Trachea

Diaphragm

**Figure 14–1**

**3.** Figure 14–2 illustrates oral cavity structures. First, correctly identify all structures provided with leader lines. Then color the structure that attaches the tongue to the floor of the mouth red; color the portions of the roof of the mouth unsupported by bone blue; color the structures that are essentially masses of lymphatic tissue yellow; and color the structure that contains the bulk of the taste buds pink.

**Figure 14–2**

**4.** Various types of glands secrete substances into the alimentary tube. Match the glands listed in Column B to the functions/locations described in Column A. Place the correct term or letter response in the answer blanks.

**Column A**

_____ 1. Produce an enzyme-poor "juice" containing mucus; found in the submucosa of the small intestine

_____ 2. Secretion includes amylase, which begins starch digestion in the mouth

_____ 3. Ducts a variety of enzymes in an alkaline fluid into the duodenum

_____ 4. Produces bile, which is transported to the duodenum via the bile duct

_____ 5. Produce hydrochloric acid and pepsinogen

**Column B**

A. Gastric glands

B. Intestinal glands

C. Liver

D. Pancreas

E. Salivary glands

**5.** Using the key choices, select the terms identified in the following descriptions by inserting the appropriate term or letter in the answer blanks.

**Key Choices**

A. Anal canal

B. Appendix

C. Colon

D. Esophagus

E. Greater omentum

F. Hard palate

G. Haustra

H. Ileocecal valve

I. Lesser omentum

J. Mesentery

K. Microvilli

L. Oral cavity

M. Parietal peritoneum

N. Peyer's patches

O. Pharynx

P. Plicae circulares

Q. Pyloric sphincter (valve)

R. Rugae

S. Small intestine

T. Soft palate

U. Stomach

V. Tongue

W. Vestibule

X. Villi

Y. Visceral peritoneum

_____ 1. Structure that suspends the small intestine from the posterior body wall

_____ 2. Fingerlike extensions of the intestinal mucosa that increase the surface area

_____ 3. Collections of lymphatic tissue found in the submucosa of the small intestine

_____ 4. Folds of the small intestine wall

_____ 5. Two anatomical regions involved in the physical breakdown of food

_____ 6. Organ that mixes food in the mouth

_____ 7. Common passage for food and air

_____ 8. Three extensions/modifications of the peritoneum

_____

_____

_____ 9. Literally a food chute; has no digestive or absorptive role

_____ 10. Folds of the stomach mucosa

_____ 11. Saclike outpocketings of the large intestine wall

_____ 12. Projections of the plasma membrane of a cell that increase the cell's surface area

_____ 13. Prevents food from moving back into the small intestine once it has entered the large intestine

_____ 14. Organ responsible for most food and water absorption

_____ 15. Organ primarily involved in water absorption and feces formation

_____ 16. Area between the teeth and lips/cheeks

_____ 17. Blind sac hanging from the initial part of the colon

_____ 18. Organ in which protein digestion begins

_____ 19. Membrane attached to the lesser curvature of the stomach

_____ 20. Organ into which the stomach empties

_____ 21. Sphincter, controlling the movement of food from the stomach into the duodenum

_____ 22. Uvula hangs from its posterior edge

_____ 23. Organ that receives pancreatic juice and bile

_____ 24. Serosa of the abdominal cavity wall

_____ 25. Region, containing two sphincters, through which feces are expelled from the body

_____ 26. Anterosuperior boundary of the oral cavity; supported by bone

**6.** Figure 14–3A is a longitudinal section of the stomach. First, use the following terms to identify the regions provided with leader lines on the figure.

| | | | |
|---|---|---|---|
| Body | Pyloric region | Greater curvature | Cardioesophageal valve |
| Fundus | Pyloric valve | Lesser curvature | |

Then select different colors for each of the following structures/areas and use them to color the coding circles and corresponding structures/areas on the figure.

○ Oblique muscle layer     ○ Longitudinal muscle layer     ○ Circular muscle layer

○ Area where rugae are visible     ○ Serosa

Figure 14–3B shows two types of secretory cells found in gastric glands. Color the hydrochloric acid–secreting cells red and color the cells that produce protein-digesting enzymes blue.

**7.** Circle the term that does not belong in each of the following groupings.

1. Nasopharynx          Esophagus          Laryngopharynx          Oropharynx

2. Villi          Plicae circulares          Rugae          Microvilli

3. Salivary glands          Pancreas          Liver          Gallbladder

4. Duodenum          Cecum          Jejunum          Ileum

5. Ascending colon          Haustra          Circular folds          Cecum

6. Mesentery          Frenulum          Greater omentum          Parietal peritoneum

7. Parotid          Sublingual          Submandibular          Palatine

8. Protein-digesting enzymes          Saliva          Intrinsic factor          HCl

9. Colon          Water absorption          Protein absorption          Vitamin B absorption

**A**

Esophagus

Duodenum

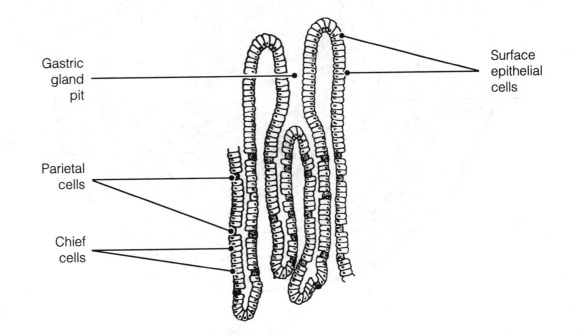

Gastric
gland
pit

Surface
epithelial
cells

Parietal
cells

Chief
cells

**B**

**Figure 14–3**

**8.** The walls of the alimentary canal have four typical layers, as illustrated in Figure 14–4. Identify each layer by placing its correct name in the space before the appropriate description. Then select different colors for each layer and use them to color the coding circles and corresponding structures on the figure. Finally, assume the figure shows a cross-sectional view of the small intestine and label the three structures provided with leader lines.

_____ ◯ 1. The secretory and absorptive layer

_____ ◯ 2. Layer composed of at least two muscle layers

_____ ◯ 3. Connective tissue layer, containing blood, lymph vessels, and nerves

_____ ◯ 4. Outermost layer of the wall; visceral peritoneum

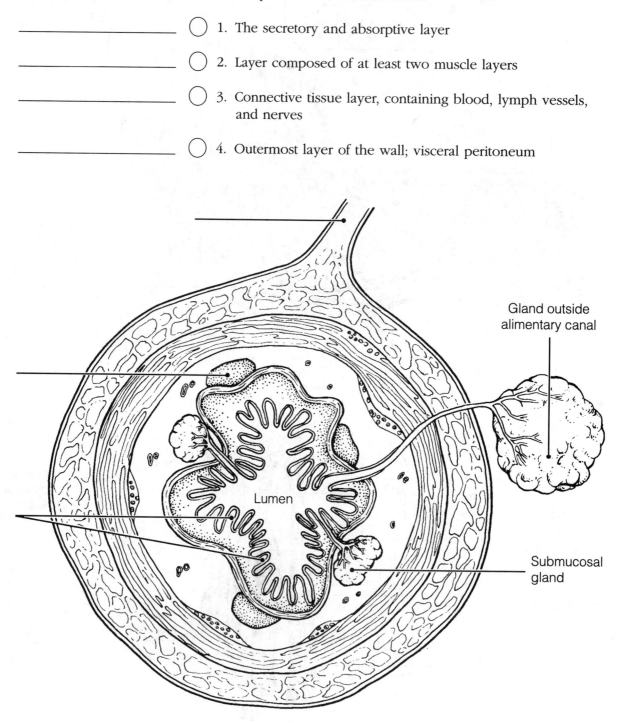

Gland outside
alimentary canal

Lumen

Submucosal
gland

**Figure 14–4**

**9.** Figure 14–5 shows three views of the small intestine. First, label the villi in views B and C and the plicae circulares in views A and B. Then select different colors for each term listed below and use them to color in the coding circles and corresponding structures in view C.

◯ Surface epithelium     ◯ Lacteal     ◯ Capillary network

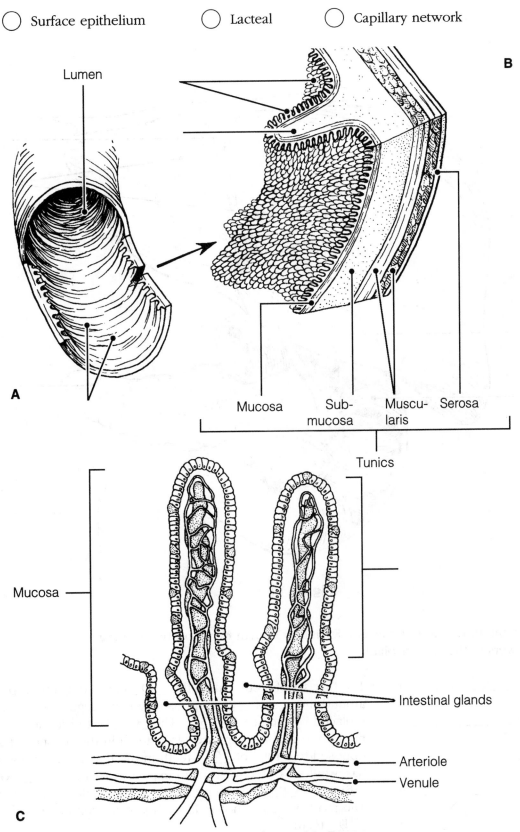

Lumen

B

A

Mucosa     Sub-mucosa     Muscu-laris     Serosa

Tunics

Mucosa

Intestinal glands

Arteriole

Venule

C

**Figure 14–5**

**10.** Three accessory organs are illustrated in Figure 14–6. Identify each of the three organs and the ligament provided with leader lines on the figure. Then select different colors for the following structures and use them to color the coding circles and the corresponding structures on the figure.

○ Common hepatic duct          ○ Bile duct

○ Cystic duct                  ○ Pancreatic duct

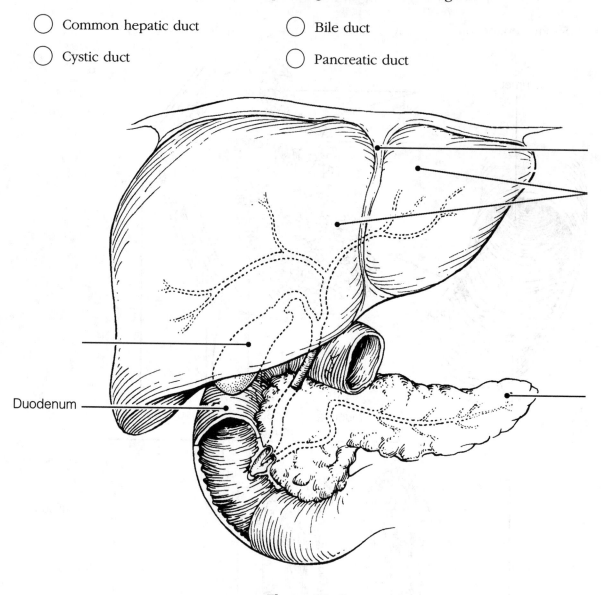

Duodenum

**Figure 14–6**

**11.** Complete the following statements referring to human dentition by inserting your answers in the answer blanks.

_____ 1.

_____ 2.

_____ 3.

_____ 4.

_____ 5.

The first set of teeth, called the __(1)__ teeth, begin to appear around the age of __(2)__ and usually have begun to be replaced by the age of __(3)__. The __(4)__ teeth are more numerous; that is, there are __(5)__ teeth in the second set as opposed to a total of __(6)__ teeth in the first set. If an adult has a full set of teeth, you can expect to find two __(7)__, one __(8)__, two __(9)__, and three __(10)__ in one side of each jaw. The most posterior molars in each jaw are commonly called __(11)__ teeth.

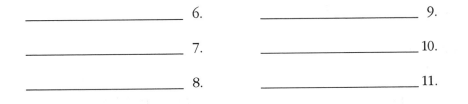

_____ 6.    _____ 9.

_____ 7.    _____ 10.

_____ 8.    _____ 11.

**12.** First, use the key choices to label the tooth diagrammed in Figure 14–7.
Second, select different colors to represent the key choices and use them
to color in the coding circles and corresponding structures in the figure.
Third, add labels to the figure to identify the crown, gingiva, and root
of the tooth.

### Key Choices

⚪ A. Cementum      ⚪ C. Enamel                          ⚪ E. Pulp

⚪ B. Dentin        ⚪ D. Periodontal membrane (ligament)

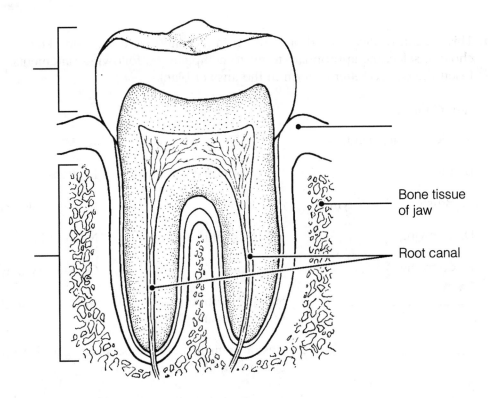

**Figure 14–7**

# FUNCTIONS OF THE DIGESTIVE SYSTEM

**13.** Match the descriptions in Column B with the appropriate terms referring to digestive processes in Column A.

| Column A | Column B |
|---|---|
| _____ 1. Ingestion | A. Transport of nutrients from lumen to blood |
| _____ 2. Propulsion | B. Enzymatic breakdown |
| _____ 3. Mechanical digestion | C. Elimination of feces |
| _____ 4. Chemical digestion | D. Eating |
| _____ 5. Absorption | E. Chewing |
| _____ 6. Defecation | F. Churning |
| | G. Includes swallowing |
| | H. Segmentation and peristalsis |

**14.** This section relates to food breakdown in the digestive tract. Using key choices, select the appropriate terms to complete the following statements. Insert the correct letter or term in the answer blanks.

*Key Choices*

| | | |
|---|---|---|
| A. Bicarbonate-rich fluid | F. HCl | K. Mucus |
| B. Bile | G. Hormonal stimulus | L. Pepsin |
| C. Brush border enzymes | H. Lipases | M. Psychological stimulus |
| D. Chewing | I. Mechanical stimulus | N. Rennin |
| E. Churning | J. Mouth | O. Salivary amylase |

_____ 1. Starch digestion begins in the mouth when __(1)__ is ducted in by the salivary glands.

_____ 2. Gastrin, which prods the stomach glands to produce more enzymes and HCl represents a __(2)__.

_____ 3. The fact that the mere thought of a relished food can make your mouth water is an example of __(3)__.

_____ 4. Many people chew gum to increase saliva formation when their mouth is dry. This type of stimulus is a __(4)__.

_____ 5. Protein foods are largely acted on in the stomach by __(5)__.

_____ 6. For the stomach protein-digesting enzymes to become active, __(6)__ is needed.

_____ 7. Since living cells of the stomach (and everywhere) are largely protein, it is amazing that they are not digested by the activity of stomach enzymes. The most important means of stomach protection is the __(7)__ it produces.

_____ 8. A milk protein–digesting enzyme found in children but uncommon in adults is __(8)__.

_____ 9. The third layer of smooth muscle found in the stomach wall allows mixing and mechanical breakdown by __(9)__.

_____ 10. Important intestinal enzymes are the __(10)__.

_____ 11. The small intestine is protected from the corrosive action of hydrochloric acid in chyme by __(11)__, which is ducted in by the pancreas.

_____ 12. The pancreas produces protein-digesting enzymes, amylase, and nucleases. It is the only important source of __(12)__.

_____ 13. A nonenzyme substance that causes fat to be dispersed into smaller globules is __(13)__.

**15.** Identify the pathologic conditions described below by using terms from the key choices. Insert the correct term or letter in the answer blanks.

*Key Choices*

A. Appendicitis     C. Diarrhea     E. Heartburn     G. Peritonitis

B. Constipation     D. Gallstones     F. Jaundice     H. Ulcer

_____ 1. Inflammation of the abdominal serosa

_____ 2. Condition resulting from the reflux of acidic gastric juice into the esophagus

_____ 3. Usually indicates liver problems or blockage of the biliary ducts

_____ 4. An erosion of the stomach or duodenal mucosa

_____ 5. Passage of watery stools

_____ 6. Causes severe epigastric pain; associated with prolonged storage of bile in the gallbladder

_____ 7. Inability to pass feces; often a result of poor bowel habits

**16.** Hormonal stimuli are important in digestive activities that occur in the stomach and small intestine. Using the key choices, identify the hormones that function as described in the following statements. Insert the correct term or letter response in the answer blanks.

**Key Choices**

A. Cholecystokinin        B. Gastrin        C. Secretin

_____ 1. These two hormones stimulate the pancreas to release its secretions.

_____ 2. This hormone stimulates increased production of gastric juice.

_____ 3. This hormone causes the gallbladder to release stored bile.

_____ 4. This hormone causes the liver to increase its output of bile.

**17.** Various types of foods are ingested in the diet and broken down to their build-ing blocks. Use the key choices to complete the following statements according to these understandings. Insert the correct term or letter in the answer blanks. In some cases, more than one choice applies.

**Key Choices**

A. Amino acids        D. Galactose        G. Maltose

B. Fatty acids        E. Glucose        H. Starch

C. Fructose        F. Lactose        I. Sucrose

_____ 
_____ 1. The building blocks of carbohydrates are monosaccharides, or simple sugars. The three common simple sugars in our diet are ___, ___, and ___.

_____ 

_____ 2. Disaccharides include ___, ___, and ___.

_____ 

_____ 

_____ 3. Protein foods must be digested to ___ before they can be absorbed.

_____ 4. Fats are broken down to two types of building blocks, ___ and glycerol.

_____ 5. Of the simple sugars, ___ is most important because it is the sugar referred to as "blood sugar."

**18.** Dietary substances capable of being absorbed are listed next. If the substance is *most often* absorbed from the digestive tract by active transport processes, put an *A* in the blank. If it is usually absorbed passively (by diffusion or osmosis), put a *P* in the blank. In addition, circle the substance that is *most likely* to be absorbed into a lacteal rather than into the capillary bed of the villus.

_____ 1. Water  _____ 3. Simple sugars  _____ 5. Electrolytes

_____ 2. Amino acids  _____ 4. Fatty acids

**19.** Complete the following statements that describe mechanisms of food mixing and movement. Insert your responses in the answer blanks.

_____ 1.

_____ 2.

_____ 3.

_____ 4.

_____ 5.

_____ 6.

_____ 7.

_____ 8.

_____ 9.

_____ 10.

_____ 11.

_____ 12.

_____ 13.

_____ 14.

_____ 15.

_____ 16.

_____ 17.

Swallowing, or __(1)__, occurs in two major phases—the __(2)__ and __(3)__. During the voluntary phase, the __(4)__ is used to push the food into the throat, and the __(5)__ rises to close off the nasal passageways. As food is moved involuntarily through the pharynx, the __(6)__ rises to ensure that its passageway is covered by the __(7)__, so that ingested substances do not enter respiratory passages. It is possible to swallow water while standing on your head because the water is carried along the esophagus involuntarily by the process of __(8)__. The pressure exerted by food on the __(9)__ valve causes it to open so that food can enter the stomach.

The two major types of movements that occur in the small intestine are __(10)__ and __(11)__. One of these movements, the __(12)__, acts to continually mix the food with digestive juices, and (strangely) also plays a major role in propelling foods along the tract. Still another type of movement seen only in the large intestine, __(13)__, occurs infrequently and acts to move feces over relatively long distances toward the anus. Presence of feces in the __(14)__ excites stretch receptors so that the __(15)__ reflex is initiated. Irritation of the gastrointestinal tract by drugs or bacteria might stimulate the __(16)__ center in the medulla, causing __(17)__, which is essentially a reverse peristalsis.

# NUTRITION AND METABOLISM

**20.** Using the key choices, identify the foodstuffs used by cells in the cellular functions described below. Insert the correct term or key letter in the answer blanks.

### Key Choices

A. Amino acids      B. Carbohydrates      C. Fats

_____ 1. The most used substance for producing the energy-rich ATP

_____ 2. Important in building myelin sheaths and cell membranes

_____ 3. Tend to be conserved by cells

_____ 4. The second most important food source for making cellular energy

_____ 5. Form insulating deposits around body organs and beneath the skin

_____ 6. Used to make the bulk of cell structure and functional substances such as enzymes

**21.** Identify the nutrients described by using the key choices. Insert the correct letter(s) in the answer blanks.

### Key Choices

A. Bread/pasta      D. Fruits      G. Starch

B. Cheese/cream      E. Meat/fish      H. Vegetables

C. Cellulose      F. Minerals      I. Vitamins

_____ 1. Examples of carbohydrate-rich *foods* in the diet.

_____ 2. Fatty foods ingested in the normal diet include _____.

_____ 3. The only important *digestible* polysaccharide.

_____ 4. An indigestible polysaccharide that aids elimination because it adds bulk to the diet is _____.

_____ 5. Protein-rich foods include _____ and _____.

_____ 6. Most examples of these nutrients, which are found largely in vegetables and fruits, are used as coenzymes.

_____ 7. Include copper, iron, and sodium.

**22.** Figure 14–8 depicts the three stages of cellular respiration. Label the figure by placing the following terms on the appropriate answer blanks. Color the diagram as suits your fancy, and then answer the questions below the figure.

| | | |
|---|---|---|
| ATP | Glucose | Mitochondrion |
| Carbon dioxide | Glycolysis | Pyruvic acid |
| Chemical energy | Electron transport chain | Water |
| Cytosol | Krebs cycle | |

**Figure 14–8**

1. Which of the oxidative phases does not require oxygen?

_____

2. Which phases do require oxygen? _____

_____

3. In what form is chemical energy transferred from the first two phases to the third phase?

_____

4. Which of the phases produces the largest amount of ATP?

_____

5. Which phase combines energetic H atoms with molecular oxygen?

_____

**23.** This section considers the process of cellular metabolism. Insert the correct word(s) from the key choices in the answer blanks.

**Key Choices**

A. ATP
B. Acetic acid
C. Acetoacetic acid
D. Acetone
E. Amino acids
F. Ammonia

G. Basal metabolic rate (BMR)
H. Carbon dioxide
I. Essential
J. Fatty acids
K. Glucose
L. Glycogen

M. Ketosis
N. Monosaccharides
O. Oxygen
P. Total metabolic rate (TMR)
Q. Urea
R. Water

_____ 1.

_____ 2.

_____ 3.

_____ 4.

_____ 5.

_____ 6.

_____ 7.

_____ 8.

_____ 9.

_____ 10.

_____ 11.

_____ 12.

The key "fuel" used by body cells is __(1)__. The cells break this fuel molecule apart piece by piece. The hydrogen removed is combined with __(2)__ to form __(3)__, while its carbon leaves the body in the form of __(4)__ gas. The importance of this process is that it provides __(5)__, a form of energy that the cells can use to power all their activities. For carbohydrates to be oxidized, or burned for energy, they must first be broken down to __(6)__. When carbohydrates are unavailable to prime the metabolic pump, intermediate products of fat metabolism such as __(7)__ and __(8)__ accumulate in the blood, causing __(9)__ and low blood pH. Amino acids are actively accumulated by cells because protein cannot be made unless all amino acid types are present. The amino acids that *must* be taken in the diet are called __(10)__ amino acids. When amino acids are oxidized to form cellular energy, their amino groups are removed and liberated as __(11)__. In the liver, this is combined with carbon dioxide to form __(12)__, which is removed from the body by the kidneys.

**24.** The liver has many functions in addition to its digestive function. Complete the following statements that elaborate on the liver's function by inserting the correct terms in the answer blanks.

_____ 1.

_____ 2.

_____ 3.

The liver is the most important metabolic organ in the body. In its metabolic role, the liver uses amino acids from the nutrient-rich hepatic portal blood to make many blood proteins such as __(1)__, which helps to hold water in the bloodstream, and __(2)__, which prevent blood loss when blood vessels are damaged. The liver also makes a steroid substance that is released to the blood. This steroid, __(3)__, has been

_____ 4.

_____ 5.

_____ 6.

_____ 7.

_____ 8.

_____ 9.

_____ 10.

_____ 11.

_____ 12.

_____ 13.

_____ 14.

_____ 15.

_____ 16.

_____ 17.

_____ 18.

_____ 19.

_____ 20.

implicated in high blood pressure and heart disease. Additionally, the liver acts to maintain homeostatic blood glucose levels. It removes glucose from the blood when blood glucose levels are high, a condition called __(4)__, and stores it as __(5)__. Then, when blood glucose levels are low, a condition called __(6)__, liver cells break down the stored carbohydrate and release glucose to the blood once again. This latter process is termed __(7)__. When the liver makes glucose from noncarbohydrate substances such as fats or proteins, the process is termed __(8)__. In addition to its processing of amino acids and sugars, the liver plays an important role in the processing of fats. Other functions of the liver include the __(9)__ of drugs and alcohol. Its __(10)__ cells protect the body by ingesting bacteria and other debris.

The liver forms small complexes called __(11)__, which are needed to transport fatty acids, fats, and cholesterol in the blood because lipids are __(12)__ in a watery medium. The function of __(13)__ is transport of cholesterol to peripheral tissues, where cells use it to construct their plasma __(14)__ or to synthesize __(15)__. The function of HDLs (high-density lipoproteins) is transport of cholesterol to the __(16)__, where it is degraded and secreted as __(17)__, which are eventually excreted. High levels of cholesterol in the plasma are of concern because of the risk of __(18)__.

Two other important functions of the liver are the storage of vitamins (such as vitamin __(19)__ needed for vision) and of the metal __(20)__ (as ferritin).

**25.** Circle the term that does not belong in each of the following groupings.

1. BMR            TMR            Rest            Postabsorptive state

2. Thyroxine        Iodine        ↓ Metabolic rate        ↑ Metabolic rate

3. Obese person        ↓ Metabolic rate        Women        Child

4. 4 Kcal/gram        Fats        Carbohydrates        Proteins

5. Radiation        Vasoconstriction        Evaporation        Vasodilation

**26.** Using the key choices, select the terms identified in the following descriptions. Insert the appropriate term(s) or letter(s) in each answer blank.

*Key Choices*

A. Blood

B. Constriction of skin blood vessels

C. Frostbite

D. Heat

E. Hyperthermia

F. Hypothalamus

G. Hypothermia

H. Perspiration

I. Radiation

J. Pyrogens

K. Shivering

_____ 1. By-product of cell metabolism

_____ 2. Means of conserving/increasing body heat

_____ 3. Means by which heat is distributed to all body tissues

_____ 4. Site of the body's thermostat

_____ 5. Chemicals released by injured tissue cells and bacteria, causing resetting of the thermostat

_____ 6. Death of cells deprived of oxygen and nutrients, resulting from withdrawal of blood from the skin circulation when the external temperature is low

_____ 7. Means of liberating excess body heat

_____ 8. Extremely low body temperature

_____ 9. Fever

# DEVELOPMENTAL ASPECTS OF THE DIGESTIVE SYSTEM

**27.** Using the key choices, select the terms identified in the following descriptions. Insert the correct term(s) or letter(s) in each answer blank.

*Key Choices*

A. Accessory organs

B. Alimentary canal

C. Appendicitis

D. Cleft palate/lip

E. Cystic fibrosis

F. Gallbladder problems

G. Gastritis

H. PKU

I. Periodontal disease

J. Peristalsis

K. Rooting

L. Sucking

M. Stomach

N. Tracheoesophageal fistula

O. Ulcers

_____ 1. Internal tubelike cavity of the embryo

_____ 2. Glands that branch out from the digestive mucosa

_____ 3. Most common congenital defect; aspiration of feeding common

_____ 4. Congenital condition characterized by a connection between digestive and respiratory passageways

_____ 5. Congenital condition in which large amounts of mucus are produced, clogging respiratory passageways and pancreatic ducts

_____ 6. Metabolic disorder characterized by an inability to properly use the amino acid phenylalanine

_____ 7. Reflex aiding the newborn baby to find the nipple

_____ 8. Vomiting is common in infants because this structure is small

_____ 9. Most common adolescent digestive system problem

_____ 10. Inflammations of the gastrointestinal tract

_____ 11. Condition of loose teeth and inflamed gums; generally seen in elderly people

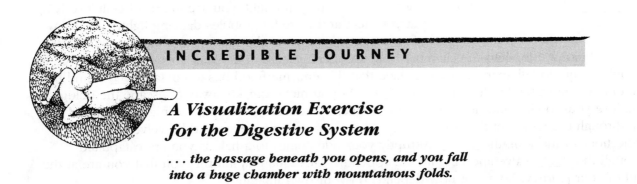

**INCREDIBLE JOURNEY**

*A Visualization Exercise for the Digestive System*

*. . . the passage beneath you opens, and you fall into a huge chamber with mountainous folds.*

**28.** Where necessary, complete statements by inserting the missing word(s) in the answer blanks.

_____ 1. In this journey you are to travel through the digestive tract as far as the appendix and then await further instructions. You

_____ 2. are miniaturized as usual and provided with a wet suit to protect you from being digested during your travels. You have a very easy entry into your host's open mouth. You look around and notice the glistening pink lining, or __(1)__, and the perfectly cared-for teeth. Within a few seconds, the lips part and you find yourself surrounded by bread. You quickly retreat to the safety of the __(2)__ between the teeth and the cheek to prevent getting chewed. From there you

_____ 3.

_____ 4.

_____ 5.

_____ 6.

_____ 7.

_____ 8.

_____ 9.

_____ 10.

_____ 11.

_____ 12.

_____ 13.

_____ 14.

_____ 15.

_____ 16.

_____ 17.

watch with fascination as a number of openings squirt fluid into the chamber, and the __(3)__ heaves and rolls, mixing the bread with the fluid.

As the bread begins to disappear, you decide that the fluid contains the enzyme __(4)__. You then walk toward the back of the oral cavity. Suddenly, you find yourself being carried along by a squeezing motion of the walls around you. The name given to this propelling motion is __(5)__. As you are carried helplessly downward, you see two openings—the __(6)__ and the __(7)__—below you. Just as you are about to straddle the solid area between them to stop your descent, the structure to your left moves quickly upward, and a trapdoor-like organ, the __(8)__, flaps over its opening. Down you go in the dark, seeing nothing. Then the passage beneath you opens, and you fall into a huge chamber with mountainous folds. Obviously, you have reached the __(9)__. The folds are very slippery, and you conclude that it must be the __(10)__ coat that you read about earlier. As you survey your surroundings, juices begin to gurgle into the chamber from pits in the "floor," and your face begins to sting and smart. You cannot seem to escape this caustic fluid and conclude that it must be very dangerous to your skin since it contains __(11)__ and __(12)__. You reach down and scoop up some of the slippery substance from the folds and smear it on your face, confident that if it can protect this organ it can protect you as well! Relieved, you begin to slide toward the organ's far exit and squeeze through the tight __(13)__ valve into the next organ. In the dim light, you see lumps of cellulose lying at your feet and large fat globules dancing lightly about. A few seconds later, your observations are interrupted by a wave of fluid pouring into the chamber from an opening high in the wall above you. The large fat globules begin to fall apart, and you decide that this enzyme flood has to contain __(14)__, and that the opening must be the duct from the __(15)__. As you move quickly away to escape the deluge, you lose your footing and find yourself on a rollercoaster ride—twisting, coiling, turning, and diving through the lumen of this active organ. As you move, you are stroked by velvety, fingerlike projections of the wall, the __(16)__. Abruptly your ride comes to a halt as you are catapulted through the __(17)__ valve and fall into the appendix. Headquarters informs you that you are at the end of your journey. Your exit now depends on your own ingenuity.

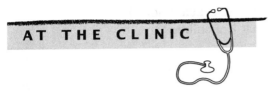

**AT THE CLINIC**

29. Mary Maroon comes to the clinic to get information on a vegetarian diet. What problems may arise when people make uninformed decisions on what to eat for a vegetarian diet? What combinations of vegetable foods will provide Mary with all the essential amino acids?

30. Mr. Ashe, a man in his mid-60s, comes to the clinic complaining of heartburn. Questioning by the clinic staff reveals that the severity of his attacks increases when he lies down after eating a heavy meal. The man is about 50 pounds overweight. What is your diagnosis? Without treatment, what conditions might develop?

31. There has been a record heat wave lately, and many elderly people are coming to the clinic complaining that they "feel poorly." In most cases, their skin is cool and clammy, and their blood pressure is low. What is their problem? What can be done to alleviate it?

32. During the same period, Bert Winchester, a construction worker, is rushed in unconscious. His skin is hot and dry, and his coworkers say that he just suddenly keeled over on the job. What is Bert's condition and how should it be handled?

33. A woman is brought to an emergency room complaining of severe pain in her left iliac region. She claims previous episodes and says that the condition is worse when she is constipated and is relieved by defecation. A large tender mass is palpated in the left iliac fossa and a barium study reveals a large number of diverticula in her descending and sigmoid colon. What are diverticula and what is believed to promote their formation? Does this woman have diverticulitis or diverticulosis? Explain.

**34.** A woman in her 50s complains of bloating, cramping, and diarrhea when she drinks milk. What is the cause of her complaint and what is a solution?

**35.** Clients are instructed not to eat before having blood tests run. How would a lab technician know if someone "cheated" and ate a fatty meal a few hours before having his blood drawn?

**36.** Zena, a teenager, has gone to the sports clinic for the past 2 years to have her fat content checked. This year, her percentage of body fat is up and tissue protein has not increased. Questioning reveals that Zena has been on crash diets four times since the last checkup, only to regain the weight (and more) each time. She also admits sheepishly that she "detests" exercise. How does cyclic dieting, accompanied by lack of exercise, cause an increase in fat and a decrease in protein?

**37.** Mrs. Rodriguez has a bleeding ulcer and has lost her appetite. She appears pale and lethargic when she comes in for a physical. She proves to be anemic and her RBCs are large and pale. What mineral supplements should be ordered?

**38.** A 21-year-old man with severe appendicitis did not seek treatment in time and died a week after his abdominal pain and fever began. Explain why appendicitis can quickly lead to death.

# CHAPTER

# 15  *The Urinary System*

Metabolism of nutrients by body cells produces various wastes such as carbon dioxide and nitrogenous wastes (creatinine, urea, and ammonia), as well as imbalances of water and essential ions. The metabolic wastes and excesses must be eliminated from the body. Essential substances are retained to ensure internal homeostasis and proper body functioning. Although several organ systems are involved in excretory processes, the urinary system bears the primary responsibility for removing nitrogenous wastes from the blood. In addition to this purely excretory function, the kidneys maintain the electrolyte, acid-base, and fluid balances of the blood. Thus, kidneys are major homeostatic organs of the body. Malfunction of the kidneys leads to a failure of homeostasis, resulting (unless corrected) in death.

Student activities in this chapter are concerned with identification of urinary system structures, and with examining urine composition and physiological processes involved in urine formation.

1. Complete the following statements by inserting your answers in the answer blanks.

_____ 1.

_____ 2.

_____ 3.

_____ 4.

_____ 5.

_____ 6.

_____ 7.

_____ 8.

_____ 9.

_____ 10.

The kidney is referred to as an excretory organ because it excretes __(1)__ wastes. It is also a major homeostatic organ because it maintains the electrolyte, __(2)__ , and __(3)__ balance of the blood. Urine is continuously formed by the __(4)__ and is routed down the __(5)__ by the mechanism of __(6)__ to a storage organ called the __(7)__. Eventually the urine is conducted to the body exterior by the __(8)__. In males, this tube-like structure is about __(9)__ inches long; in females, it is approximately __(10)__ inches long.

# KIDNEYS

## Location and Structure

2. Figure 15–1 is an anterior view of the entire urinary system. Identify and select different colors for the following organs. Use them to color the coding circles and the corresponding organs on the figure.

◯ Kidney      ◯ Bladder      ◯ Ureters      ◯ Urethra

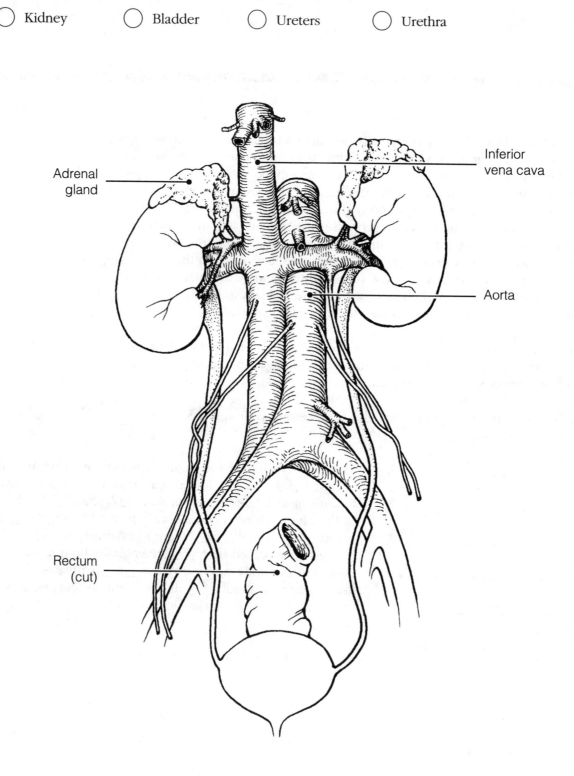

Adrenal gland

Inferior vena cava

Aorta

Rectum (cut)

**Figure 15–1**

**3.** Figure 15–2 is a longitudinal section of a kidney. First, using the correct anatomical terminology, label the following regions/structures indicated by leader lines on the figure.

- Fibrous membrane, surrounding the kidney

- Basinlike area of the kidney that is continuous with the ureter

- Cuplike extension of the pelvis that drains the apex of a pyramid

- Area of cortical tissue running through the medulla

Then, excluding the color red, select different colors to identify the following areas and structures. Then color in the coding circles and the corresponding area/structures on the figure; label these regions using the correct anatomical terms.

◯ Area of the kidney that contains the greatest proportion of nephron structures

◯ Striped-appearing structures formed primarily of collecting ducts

Finally, beginning with the renal artery, *draw in* the vascular supply to the cortex on the figure. Include and label the interlobar artery, arcuate artery, and interlobular artery. Color the vessels bright red.

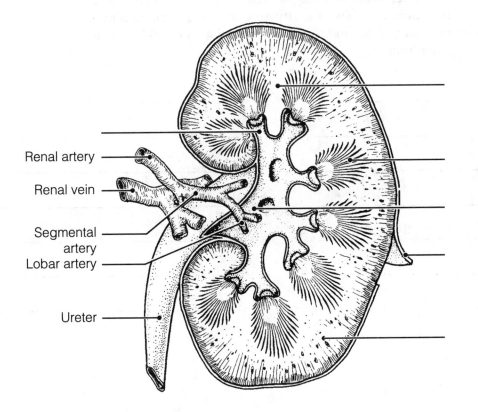

Renal artery

Renal vein

Segmental artery

Lobar artery

Ureter

**Figure 15–2**

**4.** Circle the term that does not belong in each of the following groupings.

1. Intraperitoneal        Kidney        Retroperitoneal        Superior lumbar region

2. Drains kidney        Ureter        Urethra        Renal pelvis

3. Peritubular capillaries        Reabsorption        Glomerulus        Low-pressure vessels

4. Juxtaglomerular apparatus        Distal tubule        Glomerulus        Afferent arteriole

5. Glomerulus        Peritubular capillaries        Blood vessels        Collecting duct

6. Cortical nephrons        Juxtamedullary nephrons        Cortex/medulla junction

   Long loops of Henle

7. Nephron        Proximal convoluted tubule        Distal convoluted tubule        Collecting duct

8. Medullary pyramids        Glomeruli        Renal pyramids        Collecting ducts

# Nephrons, Urine Formation, and Control of Blood Composition

**5.** Figure 15–3 is a diagram of the nephron and associated blood supply. First, match each of the numbered structures on the figure to one of the terms below the figure. Place the terms in the numbered spaces provided below. Then color the structure on the figure that contains podocytes green; the filtering apparatus red; the capillary bed that directly receives the reabsorbed substances from the tubule cells blue; the structure into which the nephron empties its urine product yellow; and the tubule area that is the primary site of tubular reabsorption orange.

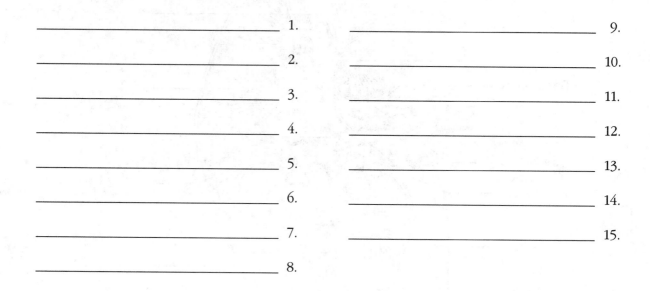

| | |
|---|---|
| _____ 1. | _____ 9. |
| _____ 2. | _____ 10. |
| _____ 3. | _____ 11. |
| _____ 4. | _____ 12. |
| _____ 5. | _____ 13. |
| _____ 6. | _____ 14. |
| _____ 7. | _____ 15. |
| _____ 8. | |

**Figure 15–3**

| | | |
|---|---|---|
| Afferent arteriole | Distal convoluted tubule | Interlobular artery |
| Arcuate artery | Efferent arteriole | Interlobular vein |
| Arcuate vein | Glomerulus | Loop of Henle |
| Glomerular capsule | Interlobar artery | Proximal convoluted tubule |
| Collecting duct | Interlobar vein | Peritubular capillaries |

**6.** Figure 15–4 is a diagram of a nephron. Add colored arrows on the figure as instructed to show the location and direction of the following processes. Add **black arrows** at the site of filtrate formation; **red arrows** at the major site of amino acid and glucose reabsorption; **green arrows** at the sites most responsive to action of ADH (show direction of water movement); **yellow arrows** at the sites most responsive to the action of aldosterone (show direction of Na⁺ movement); and **blue arrows** at the site of tubular secretion.

Then, label the proximal convoluted tubule (PCT), distal convoluted tubule (DCT), loop of Henle, glomerular capsule, and glomerulus on the figure. Also label the collecting duct (not part of the nephron) on the figure.

**Figure 15–4**

**7.** Complete the following statements by inserting your answers in the answer blanks.

_____ 1.

_____ 2.

_____ 3.

_____ 4.

_____ 5.

_____ 6.

_____ 7.

_____ 8.

_____ 9.

_____ 10.

_____ 11.

_____ 12.

_____ 13.

_____ 14.

_____ 15.

_____ 16.

The glomerulus is a unique high-pressure capillary bed, because the __(1)__ arteriole feeding it is larger in diameter than the __(2)__ arteriole draining the bed. Glomerular filtrate is very similar to __(3)__, but it has fewer proteins. Mechanisms of tubular reabsorption include __(4)__ and __(5)__. As an aid for the reabsorption process, the cells of the proximal convoluted tubule have dense __(6)__ on their luminal surface, which increase the surface area dramatically. Other than reabsorption, an important tubule function is __(7)__, which is important for ridding the body of substances not already in the filtrate. Blood composition depends on __(8)__, __(9)__, and __(10)__. In a day's time, 180 liters of blood plasma are filtered into the kidney tubules, but only about __(11)__ liters of urine are actually produced. __(12)__ is responsible for the normal yellow color of urine. The three major nitrogenous wastes found in the blood, which must be disposed of, are __(13)__, __(14)__, and __(15)__. The kidneys are the final "judges" of how much water is to be lost from the body. When water loss via vaporization from the __(16)__ or __(17)__ from the skin is excessive, urine output __(18)__. If the kidneys become nonfunctional, __(19)__ is used to cleanse the blood of impurities.

_____ 17

_____ 18.

_____ 19.

**8.** Circle the term that does not belong in each of the following groupings. (BP = Blood pressure.)

1. Hypothalamus          ADH          Aldosterone          Osmoreceptors

2. Glomerulus          Secretion          Filtration          ↑ BP

3. Aldosterone          ↑ $Na^+$ reabsorption          ↑ $K^+$ reabsorption          ↑ BP

4. ADH          ↓ BP          ↑ Blood volume          ↑ Water reabsorption

5. ↓ Aldosterone          Edema          ↓ Blood volume          ↓ $K^+$ retention

6. ↓ Urine pH          ↑ $H^+$ in urine          ↑ $HCO_3^-$ in urine          ↑ Ketones

**9.** Decide whether the following conditions would cause urine to become more acidic or more basic. If more acidic, insert an *A* in the blank; if more basic, insert a *B* in the blank.

\_\_\_\_\_ 1. Protein-rich diet      \_\_\_\_\_ 4. Diabetes mellitus

\_\_\_\_\_ 2. Bacterial infection      \_\_\_\_\_ 5. Vegetarian diet

\_\_\_\_\_ 3. Starvation

**10.** Decide whether the following conditions would result in an increase or decrease in urine specific gravity. Insert *I* in the answer blank to indicate an increase and *D* to indicate a decrease.

\_\_\_\_\_ 1. Drinking excessive fluids      \_\_\_\_\_ 4. Using diuretics

\_\_\_\_\_ 2. Chronic renal failure      \_\_\_\_\_ 5. Limited fluid intake

\_\_\_\_\_ 3. Pyelonephritis      \_\_\_\_\_ 6. Fever

**11.** Assuming *normal* conditions, note whether each of the following substances would be (G) in greater concentration in the urine than in the glomerular filtrate, (L) in lesser concentration in the urine than in the glomerular filtrate, or (A) absent in both urine and glomerular filtrate. Place the correct letter in the answer blanks.

\_\_\_\_\_ 1. Water      \_\_\_\_\_ 5. Glucose      \_\_\_\_\_ 9. Potassium ions

\_\_\_\_\_ 2. Urea      \_\_\_\_\_ 6. Albumin      \_\_\_\_\_ 10. Red blood cells

\_\_\_\_\_ 3. Uric acid      \_\_\_\_\_ 7. Creatinine      \_\_\_\_\_ 11. Sodium ions

\_\_\_\_\_ 4. Pus (white blood cells)      \_\_\_\_\_ 8. Hydrogen ions      \_\_\_\_\_ 12. Amino acids

**12.** Several specific terms are used to indicate the presence of abnormal urine constituents. Identify each of the following abnormalities by inserting the term that names the condition in the spaces provided. Then for each condition, provide one possible cause of the condition in the remaining spaces.

1. Presence of red blood cells: _____. Cause: _____.

2. Presence of ketones: _____. Cause: _____.

3. Presence of albumin: _____. Cause: _____.

4. Presence of pus: _____. Cause: _____.

5. Presence of bile: _____. Cause: _____.

6. Presence of "sand": _____. Cause: _____.

7. Presence of glucose: _____. Cause: _____.

**13.** Glucose and albumin are both normally absent from urine, but the reason for their exclusion differs. Respond to the following questions in the spaces provided.

1. Explain the reason for the absence of glucose in urine. _____

_____

2. Explain the reason for the absence of albumin in urine. _____

_____

**14.** By what three methods is $H^+$ concentration in body fluids regulated? Also give the approximate time for each method to respond to pH changes.

1. _____

2. _____

3. _____

**15.** Circle the term that does not belong in each of the following groupings. (ECF = Extracellular fluid compartment)

1. Female adult        Male adult        About 50% water        Less muscle

2. Obese adult        Lean adult        Less body water        More adipose tissue

3. ECF        Interstitial fluid        Intracellular fluid        Plasma

4. Electric charge        Nonelectrolyte        Ions        Conducts a current

5. ↑ Water output        ↓ $Na^+$ concentration        ↑ ADH        ↓ ADH

6. Aldosterone        ↑ $Na^+$ reabsorption        ↑ $K^+$ reabsorption        ↑ BP

# URETERS, URINARY BLADDER, AND URETHRA

**16.** Circle the term that does not belong in each of the following groupings.

1. Bladder        Kidney        Transitional epithelium        Detrusor muscle

2. Trigone        Ureter openings        Urethral opening        Bladder        Forms urine

3. Surrounded by prostate gland        Contains internal and external sphincters

Continuous with renal pelvis        Urethra

**17.** Using the key choices, identify the structures that best fit the following descriptions. Insert the correct term(s) or corresponding letter(s) in the answer blanks.

*Key Choices*

A. Bladder      B. Urethra      C. Ureter

_____ 1. Drains the bladder

_____ 2. Storage area for urine

_____ 3. Contains the trigone

_____ 4. In males has prostatic, membranous, and spongy parts

_____ 5. Conducts urine by peristalsis

_____ 6. Substantially longer in males than in females

_____ 7. A common site of "trapped" renal calculi

_____ 8. Contains transitional epithelium

_____ 9. Also transports sperm in males

**18.** Complete the following statements by inserting your answers in the answer blanks.

_____ 1.

_____ 2.

_____ 3.

_____ 4.

_____ 5.

_____ 6.

_____ 7.

_____ 8.

_____ 9.

_____ 10.

_____ 11.

_____ 12.      _____ 13.

Another term that means voiding or emptying of the bladder is __(1)__. Voiding has both voluntary and involuntary aspects. As urine accumulates in the bladder, __(2)__ are activated. This results in a reflex that causes the muscular wall of the bladder to __(3)__, and urine is forced past the __(4)__ sphincter. The more distal __(5)__ sphincter is controlled __(6)__; thus an individual can temporarily postpone emptying the bladder until it has accumulated __(7)__ ml of urine. __(8)__ is a condition in which voiding cannot be voluntarily controlled. It is normal in __(9)__, because nervous control of the voluntary sphincter has not been achieved. Other conditions that might result in an inability to control the sphincter include __(10)__ and __(11)__. __(12)__ is essentially the opposite of incontinence and often is a problem in elderly men due to __(13)__ enlargement.

**19.** Match the terms in Column B with the appropriate descriptions provided
in Column A. Insert the correct term or letter in the answer blanks.

**Column A**

**Column B**

_____ 1. Inflammatory condition common in
women with poor toileting habits

_____ 2. Backup of urine into the kidney; often
a result of a blockage in the urinary tract

_____ 3. Toxic condition due to renal failure

_____ 4. Inflammation of a kidney

_____ 5. A condition in which excessive amounts
of urine are produced because of a deficiency of antidiuretic hormone (ADH)

_____ 6. Dropping of the kidney to a more inferior position in the
abdomen; may result from a rapid weight loss that decreases
the fatty cushion surrounding the kidney

A. Cystitis

B. Diabetes insipidus

C. Hydronephrosis

D. Ptosis

E. Pyelonephritis

F. Uremia

# DEVELOPMENTAL ASPECTS OF THE URINARY SYSTEM

**20.** Complete the following statements by inserting your responses in the
answer blanks.

_____ 1.

_____ 2.

_____ 3.

_____ 4.

_____ 5.

_____ 6.

_____ 7.

_____ 8.

_____ 9.

_____ 10.

_____ 11.

_____ 12.

Three separate sets of renal tubules develop in the embryo;
however, embryonic nitrogenous wastes are actually disposed
of by the __(1)__. A congenital condition typified by blisterlike
sacs in the kidneys is __(2)__. __(3)__ is a congenital condition
seen in __(4)__, when the urethral opening is located ventrally
on the penis. A newborn baby voids frequently, which
reflects its small __(5)__. Daytime control of the voluntary urethral sphincter is usually achieved by approximately __(6)__
months. Urinary tract infections are fairly common and not
usually severe with proper medical treatment. A particularly
problematic condition, called __(7)__, may result later in life as
a sequel to childhood streptococcus infection. In this disease,
the renal filters become clogged with __(8)__ complexes, urine
output decreases, and __(9)__ and __(10)__ begin to appear in
the urine. In old age, progressive __(11)__ of the renal blood
vessels results in the death of __(12)__ cells. The loss of bladder
tone leads to __(13)__ and __(14)__ and is particularly
troublesome to elderly people.

_____ 13.

_____ 14.

*A Visualization Exercise for the Urinary System*

**You see the kidney looming brownish red through the artery wall.**

**21.** Where necessary, complete statements by inserting the missing word(s) in the answer blanks.

_____ 1.

_____ 2.

_____ 3.

_____ 4.

_____ 5.

_____ 6.

_____ 7.

_____ 8.

_____ 9.

_____ 10.

For your journey through the urinary system, you must be made small enough to filter through the filtration membrane from the bloodstream into a renal __(1)__ . You will be injected into the subclavian vein and must pass through the heart before entering the arterial circulation. As you travel through the systemic circulation, you have at least two minutes to relax before reaching the __(2)__ artery, feeding a kidney. You see the kidney looming brownish red through the artery wall. Once you have entered the kidney, the blood vessel conduits become increasingly smaller until you finally reach the __(3)__ arteriole, feeding into the filtering device, or __(4)__ . Once in the filter, you maneuver yourself so that you are directly in front of a pore. Within a fraction of a second, you are swept across the filtration membrane into the __(5)__ part of the nephron. Drifting along, you lower the specimen cup to gather your first filtrate sample for testing. You study the readout from the sample and note that it is very similar in composition to __(6)__ with one exception: There are essentially no __(7)__ . Your next sample doesn't have to be collected until you reach the "hairpin," or, using the proper terminology, the __(8)__ part of the tubule. As you continue your journey, you notice that the tubule cells have dense fingerlike projections extending from their surfaces into the lumen of the tubule. These are __(9)__ , which increase the surface area of tubules because this portion of the tubule is very active in the process of __(10)__ . Soon you collect your second sample, and then later, in the distal convoluted tubule, your third sample. When you read the computer's summary of the third sample, you make the following notes in your register.

_____ 11.

_____ 12.

_____ 13.

_____ 14.

_____ 15.

_____ 16.

_____ 17.

_____ 18.

_____ 19.

_____ 20.

_____ 21.

_____ 22.

_____ 23.

_____ 24.

- Virtually no nutrients such as  (11)  and  (12)  are left in the filtrate.

- The pH is acid, 6.0. This is quite a change from the pH of  (13)  recorded for the newly formed filtrate.

- There is a much higher concentration of  (14)  wastes here.

- There are many fewer  (15)  ions but more of the  (16)  ions noted.

- Color of the filtrate is yellow, indicating a high relative concentration of the pigment  (17) .

Gradually you become aware that you are moving along much more quickly. You see that the water level has dropped dramatically and that the stream is turbulent and rushing. As you notice this, you realize that the hormone  (18)  must have been released recently to cause this water drop. You take an abrupt right turn and then drop straight downward. You realize that you must be in a  (19) . Within a few seconds, you are in what appears to be a large tranquil sea with an ebbing tide toward a darkened area at the far shore. You drift toward the darkened area, confident that you are in the kidney  (20) . As you reach and enter the dark tubelike structure seen from the opposite shore, your progress becomes rhythmic—something like being squeezed through a sausage skin. Then you realize that your progress is being regulated by the process of  (21) . Suddenly, you free-fall and land in the previously stored  (22)  in the bladder, where the air is very close. Soon the walls of the bladder begin to gyrate, and you realize you are witnessing a  (23)  reflex. In a moment, you are propelled out of the bladder and through the  (24)  to exit from your host.

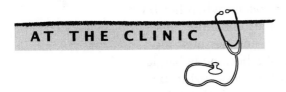

## AT THE CLINIC

22. A man was admitted to the hospital after being trampled by his horse. He received crushing blows to his lower back, on both sides. He is in considerable pain, and his chart shows a urine output of 70 ml in the last 24 hours. What is this specific symptom called? What will be required if the renal effects of his trauma persist?

**23.** Four-year-old Eddie is a chronic bedwetter. He wets the bed nearly every night. What might explain his problem?

**24.** If a tumor of the glucocorticoid-secreting cells of the adrenal cortex crowds out the cells that produce aldosterone, what is the likely effect on urine composition and volume?

**25.** Jimmy has been stressed out lately as he has been juggling two jobs while taking classes at a local college. He appears at the clinic complaining of a pounding headache. Tests show that he has high blood pressure and his cortiscosteroid levels are elevated. What is the relationship between his stress and his signs and symptoms?

**26.** Mr. O'Toole is very drunk when he is brought to the emergency room after falling down City Hall steps during a political rally. He is complaining about his "cotton mouth." Knowing that alcohol inhibits ADH's action, you explain to him why his mouth is so dry. What do you tell him?

**27.** Mrs. Rodriques is breathing rapidly and is slurring her speech when her husband calls the clinic in a panic. Shortly after, she becomes comatose. Tests show that her blood glucose and ketone levels are high and her husband said that she was urinating every few minutes before she became lethargic. What is Mrs. Rodriques's problem? Would you expect her blood pH to be acidic or alkaline? What is the significance of her rapid breathing? Are her kidneys reabsorbing or secreting bicarbonate ions during this crisis?

# 16 *The Reproductive System*

The biological function of the reproductive system is to produce offspring. The essential organs are those producing the germ cells (testes in males and ovaries in females). The male manufactures sperm and delivers them to the female's reproductive tract. The female, in turn, produces eggs. If the time is suitable, the egg and sperm fuse, producing a fertilized egg, which is the first cell of the new individual. Once fertilization has occurred, the female uterus protects and nurtures the developing embryo.

In this chapter, student activities concern the structures of the male and female reproductive systems, germ cell formation, the menstrual cycle, and embryonic development.

## ANATOMY OF THE MALE REPRODUCTIVE SYSTEM

1. Using the following terms, trace the pathway of sperm from the testis to the urethra: rete testis, epididymis, seminiferous tubule, ductus deferens. List the terms in the proper order in the spaces provided.

   _____ → _____ → _____ → _____ .

2. How do the scrotal muscles help maintain temperature homeostasis of the testes?

   _____

   _____

   _____

**3.** Using the key choices, select the terms identified in the following descriptions. Insert the appropriate term(s) or corresponding letter(s) in the answer blanks.

***Key Choices***

| | | |
|---|---|---|
| A. Bulbourethral glands | E. Penis | I. Scrotum |
| B. Epididymis | F. Prepuce | J. Spermatic cord |
| C. Ductus deferens | G. Prostate gland | K. Testes |
| D. Glans penis | H. Seminal vesicles | L. Urethra |

_____ 1. Organ that delivers semen to the female reproductive tract

_____ 2. Site of testosterone production

_____ 3. Passageway from the epididymis to the ejaculatory duct

_____ 4. Conveys both sperm and urine down the length of the penis

_____ 5. Organs that contribute to the formation of semen

_____ 6. External skin sac that houses the testes

_____ 7. Tubular storage site for sperm; hugs the lateral aspect of the testes

_____ 8. Cuff of skin encircling the glans penis

_____ 9. Surrounds the urethra at the base of the bladder; produces a milky fluid

_____ 10. Produces over half of the seminal fluid

_____ 11. Produces a lubricating mucus that cleanses the urethra

_____ 12. Connective tissue sheath enclosing the ductus deferens, blood vessels, and nerves.

**4.** Figure 16–1 is a sagittal view of the male reproductive structures. First, identify the following organs on the figure by placing each term at the end of the appropriate leader line.

| | | |
|---|---|---|
| Bulbourethral gland | Epididymis | Scrotum |
| Ductus deferens | Prepuce | Seminal vesicle |
| Glans penis | Prostate gland | Testis |
| Ejaculatory duct | Urethra | |

Next, select different colors for the structures that correspond to the following descriptions, and color in the coding circles and the corresponding structures on the figure.

○ Spongy tissue that is engorged with blood during erection

○ Portion of the duct system that also serves the urinary system

○ Structure that provides the ideal temperature conditions for sperm formation

○ Structure removed in circumcision

○ Gland whose secretion contains sugar to nourish sperm

○ Structure cut or cauterized during a vasectomy

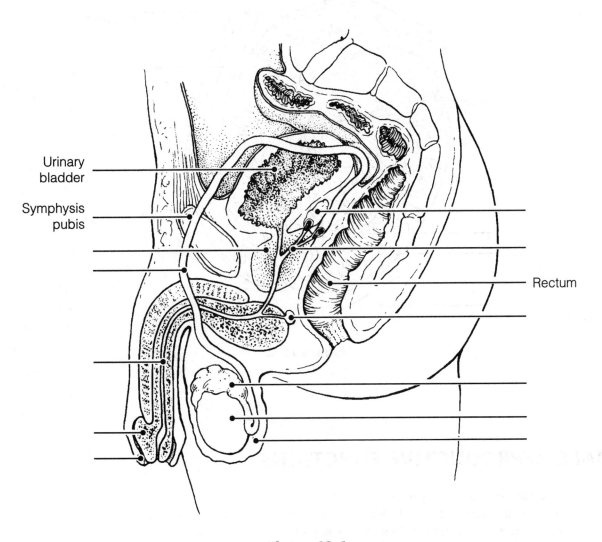

**Figure 16–1**

5. Figure 16–2 is a longitudinal section of a testis. First, select different colors for the structures that correspond to the following descriptions. Then color the coding circles and color and label the corresponding structures on the figure. *Complete the labeling* of the figure by adding the following terms: lobule, rete testis, and septum.

◯ Site(s) of spermatogenesis

◯ Tubular structure in which sperm mature and become motile

◯ Fibrous coat protecting the testis

Ductus deferens

**Figure 16–2**

# MALE REPRODUCTIVE FUNCTIONS

6. This section considers the process of sperm production in the testis. Figure 16–3 is a cross-sectional view of a seminiferous tubule in which spermatogenesis is occurring. First, using key choices, select the terms identified in the following descriptions.

*Key Choices*

A.    Follicle-stimulating hormone (FSH)    E. ◯ Sperm

B. ◯ Primary spermatocyte                  F. ◯ Spermatid

C. ◯ Secondary spermatocyte                G.    Testosterone

D. ◯ Spermatogonium

_____ 1. Primitive stem cell

_____ 2. Contain 23 chromosomes (3 answers)

_____ and _____

_____ 3. Product of meiosis I

_____ 4. Product of meiosis II

_____ 5. Functional motile gamete

_____ 6. Two hormones necessary for sperm production

_____

Then label the cells with leader lines. Select different colors for the cell types with color-coding circles listed in the key choices and color in the coding circles and corresponding structures on the figure. In addition, label and color the cells that produce testosterone.

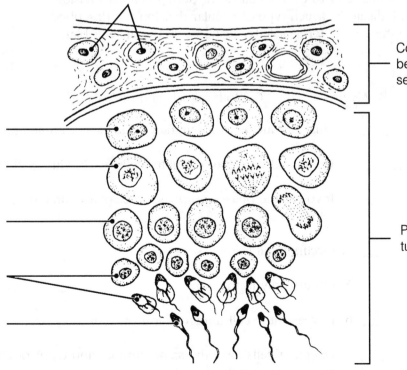

Connective tissue area between adjacent seminiferous tubules

Portion of seminiferous tubule wall

**Figure 16–3**

7. Figure 16–4 illustrates a single sperm. On the figure, bracket and label the head and the midpiece and circle and label the tail. Select different colors for the structures that correspond to the following descriptions. Color the coding circles and corresponding structures on the figure. Then label the structures, using correct terminology.

○ The DNA-containing area

○ The enzyme-containing sac that aids sperm penetration of the egg

○ Metabolically active organelles that provide ATP to energize sperm movement

**Figure 16–4**

8. The following statements refer to events that occur during cellular division. Using key choices, indicate in which type of cellular division the described events occur. Place the correct term or letter response in the answer blanks.

*Key Choices*

A. Mitosis          B. Meiosis          C. Both mitosis and meiosis

_____ 1. Final product is two daughter cells, each with 46 chromosomes

_____ 2. Final product is four daughter cells, each with 23 chromosomes

_____ 3. Involves the phases prophase, metaphase, anaphase, and telophase

_____ 4. Occurs in all body tissues

_____ 5. Occurs only in the gonads

_____ 6. Increases the cell number for growth and repair

_____ 7. Daughter cells have the same number and types of chromosomes as the mother cell

_____ 8. Daughter cells are different from the mother cell in their chromosomal makeup

_____ 9. Chromosomes are replicated before the division process begins

_____ 10. Provides cells for the reproduction of offspring

_____ 11. Consists of two consecutive divisions of the nucleus; chromosomes are not replicated before the second division

**9.** Name four of the male secondary sex characteristics. Insert your answers on the lines provided.

_____

_____

_____

_____

# ANATOMY OF THE FEMALE REPRODUCTIVE SYSTEM

**10.** Identify the female structures described by inserting your responses in the answer blanks.

_____ 1. Chamber that houses the developing fetus

_____ 2. Canal that receives the penis during sexual intercourse

_____ 3. Usual site of fertilization

_____ 4. Erects during sexual stimulation

_____ 5. Duct through which the ovum travels to reach the uterus

_____ 6. Membrane that partially closes the vaginal canal

_____ 7. Primary female reproductive organ

_____ 8. Move to create fluid currents to draw the ovulated egg into the uterine (fallopian) tube

**11.** Figure 16–5 is a sagittal view of the female reproductive organs. First, label all structures on the figure provided with leader lines. Then select different colors for the following structures, and use them to color the coding circles and corresponding structures on the figure.

○ Lining of the uterus, endometrium

○ Muscular layer of the uterus, myometrium

○ Pathway along which an egg travels from the time of its release to its implantation

○ Ligament helping to anchor the uterus

○ Structure producing female hormones and gametes

**Figure 16–5**

**12.** Figure 16–6 is a ventral view of the female external genitalia. Label the clitoris, labia minora, urethral orifice, hymen, mons pubis, and vaginal orifice on the figure. These structures are indicated with leader lines. Then color the homologue of the male penis blue, color the membrane that partially obstructs the vagina yellow, and color the distal end of the birth canal red.

Labia majora
(spread)

Anus

**Figure 16–6**

# FEMALE REPRODUCTIVE FUNCTIONS AND CYCLES

**13.** Using the key choices, identify the cell type you would expect to find in the following structures. Insert the correct term or letter response in the answer blanks.

*Key Choices*

A. Oogonium          C. Secondary oocyte

B. Primary oocyte    D. Ovum

_____ 1. Forming part of the primary follicle in the ovary

_____ 2. In the uterine tube before fertilization

_____ 3. In the mature, or Graafian, follicle of the ovary

_____ 4. In the uterine tube shortly after sperm penetration

**14.** Figure 16–7 is a sectional view of the ovary. First, identify all structures indicated with leader lines on the figure. Second, select different colors for the following structures, and use them to color the coding circles and corresponding structures on the figure.

◯ Cells that produce estrogen

◯ Glandular structure that produces progesterone

◯ All oocytes

Cell type, specific

Event A

**Figure 16–7**

Third, in the space provided, name the event depicted as "Event A" on the figure. _____

Fourth, answer the following questions by inserting your answers in the spaces provided.

1. Are there any oogonia in a mature female's ovary? _____

2. Into what area is the ovulated cell released? _____

3. When is a mature ovum (egg) produced in humans? _____

4. What structure in the ovary becomes a corpus luteum? _____

5. What are the four final cell types produced by oogenesis in the female? (Name the cell

   type and number of each.) _____

6. How does this compare with the final product of spermatogenesis in males? _____

_____

7. What happens to the tiny cells nearly devoid of cytoplasm ultimately produced during

oogenesis? _____

8. Why? _____

9. What name is given to the period of a woman's life when her ovaries begin to become

nonfunctional? _____

**15.** What is the significance of the fact that the uterine tubes are not structurally
continuous with the ovaries? Address this question from both reproductive
and health aspects.

_____

_____

_____

**16.** The following statements deal with anterior pituitary and ovarian hormonal
interrelationships. Name the hormone(s) described in each statement.
Place your answers in the answer blanks.

_____ 1. Promotes growth of ovarian follicles and production of estrogen

_____ 2. Triggers ovulation

_____ 3. Inhibit follicle-stimulating hormone (FSH) release by the anterior
pituitary

_____ 4. Stimulates luteinizing hormone (LH) release by the anterior
pituitary

_____ 5. Converts the ruptured follicle into a corpus luteum and causes it
to produce progesterone and estrogen

_____ 6. Maintains the hormonal production of the corpus luteum

**17.** Name four of the secondary sex characteristics of females. Place your answers
in the spaces provided.

_____        _____

_____        _____

**18.** Use the key choices to identify the ovarian hormone(s) responsible for the following events. Insert the correct term(s) or letter(s) in the answer blanks.

*Key Choices*

A. Estrogens        B. Progesterone

_____ 1. Lack of this (these) causes the blood vessels to kink and the endometrium to slough off (menses)

_____ 2. Causes the endometrial glands to begin the secretion of nutrients

_____ 3. The endometrium is repaired and grows thick and velvety

_____ 4. Maintains the myometrium in an inactive state if implantation of an embryo has occurred

_____ 5. Glands are formed in the endometrium

_____ 6. Responsible for the secondary sex characteristics of females

**19.** The following exercise refers to Figure 16–8 A–D.

On Figure 16–8A, the blood levels of two gonadotropic hormones (FSH and LH) of the anterior pituitary are indicated. Identify each hormone by appropriately labeling the blood level lines on the figure. Then select different colors for each of the blood level lines and color them in on the figure.

On Figure 16–8B, identify the blood level lines for the ovarian hormones, estrogens and progesterone. Then select different colors for each blood level line, and color them in on the figure.

On Figure 16–8C, select different colors for the following structures and use them to color in the coding circles and corresponding structures in the figure.

○ Primary follicle        ○ Growing follicle

○ Vesicular follicle        ○ Corpus luteum

○ Ovulating follicle

On Figure 16–8D, identify the endometrial changes occurring during the menstrual cycle by color-coding and coloring the areas depicting the three phases of that cycle.

○ Secretory stage        ○ Menses        ○ Proliferative stage

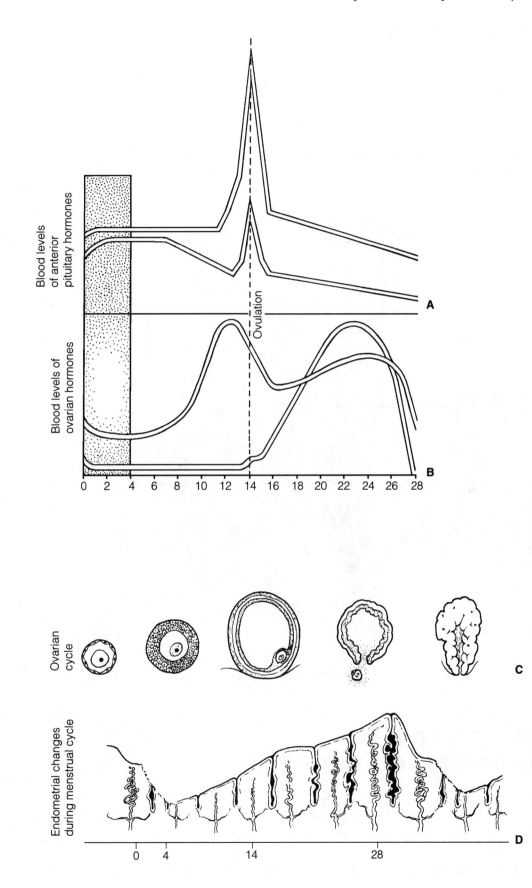

**Figure 16-8**

# MAMMARY GLANDS

**20.** Figure 16–9 is a sagittal section of a breast. First, use the following terms to correctly label all structures provided with leader lines on the figure.

Alveolar glands        Areola        Lactiferous duct        Nipple

Then color the structures that produce milk blue and color the fatty tissue of the breast yellow.

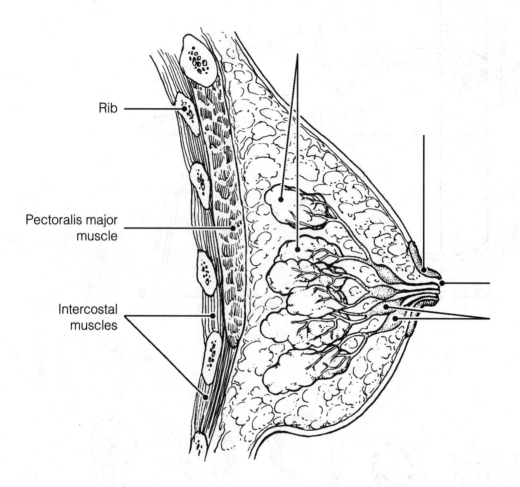

Rib

Pectoralis major muscle

Intercostal muscles

**Figure 16–9**

# SURVEY OF PREGNANCY AND EMBRYONIC DEVELOPMENT

**21.** Relative to events of sperm penetration:

1. What portion of the sperm actually enters the oocyte? _____

2. What is the functional importance of the acrosomal reaction? _____

_____

_____

**22.** Figure 16–10 depicts early embryonic events. Identify the following,
referring to the figure. Place your answers in the spaces provided.

1. Event A _____

2. Cell resulting from event A _____

3. Process B _____

4. Embryonic structure B₁ _____

5. Completed process C _____

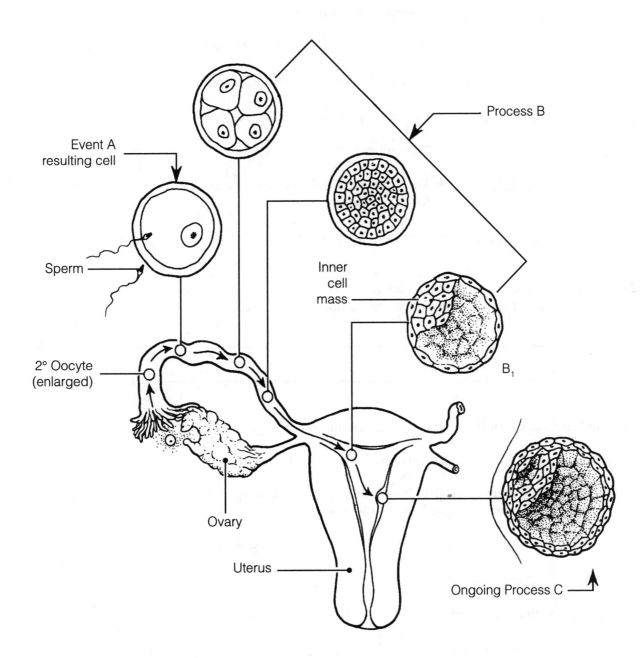

**Figure 16–10**

**23.** Using the key choices, select the terms that are identified in the following descriptions. Insert the correct term or letter response in the answer blanks.

*Key Choices*

A. Amnion          D. Fertilization          G. Umbilical cord

B. Chorionic villi  E. Fetus                  H. Zygote

C. Endometrium     F. Placenta

_____ 1. The fertilized egg

_____ 2. Secretes estrogen and progesterone to maintain the pregnancy

_____ 3. Cooperate to form the placenta

_____ 4. Fluid-filled sac, surrounding the developing embryo/fetus

_____ 5. Attaches the embryo to the placenta

_____ 6. Fingerlike projections of the blastocyst

_____ 7. The embryo after 8 weeks

_____ 8. The organ that delivers nutrients to and disposes of wastes for the fetus

_____ 9. Event leading to combination of ovum and sperm "genes"

**24.** Explain why the corpus luteum does not stop producing its hormones (estrogens and progesterone) when fertilization has occurred.

_____

_____

**25.** The first "tissues" of the embryo's body are the primary germ layers:

A. Ectoderm          B. Mesoderm          C. Endoderm

Indicate which germ layer gives rise to each of the following structures by placing the corresponding letter in the answer blank.

_____ 1. Heart and blood vessels        _____ 5. Skin epidermis

_____ 2. Digestive system mucosa        _____ 6. Bones

_____ 3. Brain and spinal cord          _____ 7. Respiratory system mucosa

_____ 4. Skeletal muscles               _____ 8. Liver and pancreas

**26.** What two hormones are essential to initiate labor in humans?

_____

**27.** 1. What hormone is responsible for milk production? _____

2. For milk ejection? _____

**28.** A pregnant woman undergoes numerous changes during her pregnancy—anatomical, metabolic, and physiological. Several such possibilities are listed below. Check (✓) all that are commonly experienced during pregnancy.

| | | | |
|---|---|---|---|
| _____ | 1. Diaphragm descent is impaired | _____ | 7. Metabolic rate declines |
| _____ | 2. Breasts decline in size | _____ | 8. Increased mobility of GI tract |
| _____ | 3. Pelvic ligaments are relaxed by relaxin | _____ | 9. Blood volume and cardiac output increase |
| _____ | 4. Vital capacity decreases | _____ | 10. Nausea, heartburn, constipation |
| _____ | 5. Lordosis | _____ | 11. Dyspnea may occur |
| _____ | 6. Blood pressure and pulse rates decline | _____ | 12. Urgency and stress incontinence |

**29.** What are Braxton Hicks contractions, and why do they occur?

_____

_____

_____

_____

**30.** Name the three phases of parturition, and briefly describe each phase.

1. _____

_____

2. _____

_____

3. _____

_____

**31.** The very simple flowchart in Figure 16–11 illustrates the sequence of events that occur during labor. Complete the flowchart by filling in the missing terms in the boxes. Use color as desired.

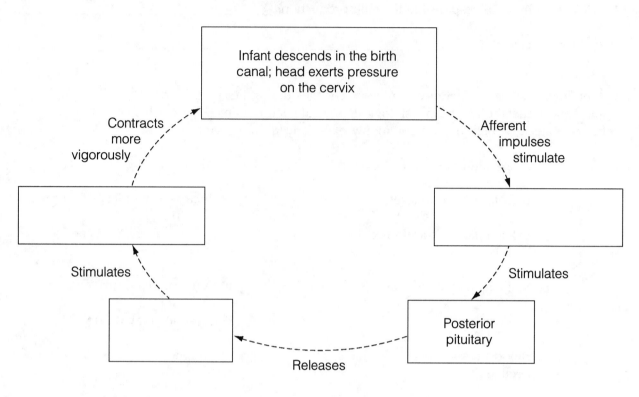

**Figure 16–11**

**32.** How long will the cycle illustrated in Figure 16–11 continue to occur?

_____

**33.** Labor is an example of a positive feedback mechanism. What does that mean?

_____

_____

_____

# DEVELOPMENTAL ASPECTS
# OF THE REPRODUCTIVE SYSTEM

**34.** Complete the following statements by inserting your responses in the answer blanks.

_____ 1.

_____ 2.

_____ 3.

_____ 4.

_____ 5.

_____ 6.

_____ 7.

_____ 8.

_____ 9.

_____ 10.

_____ 11.

_____ 12.

_____ 13.

_____ 14.

_____ 15.

_____ 16.

_____ 17.

_____ 18.

_____ 19.

_____ 20.

_____ 21.

_____ 22.

A male embryo has __(1)__ sex chromosomes, whereas a female has __(2)__ . During early development, the reproductive structures of both sexes are identical, but by the eighth week, __(3)__ begin to form if testosterone is present. In the absence of testosterone, __(4)__ form. The testes of a male fetus descend to the scrotum shortly before birth. If this does not occur, the resulting condition is called __(5)__ .

The most common problem affecting the reproductive organs of women are infections, particularly __(6)__ , __(7)__ , and __(8)__ . When the entire pelvis is inflamed, the condition is called __(9)__ . Most male problems involve inflammations, resulting from __(10)__ . A leading cause of cancer death in adult women is cancer of the __(11)__ ; the second most common female reproductive system cancer is cancer of the __(12)__ . Thus a yearly __(13)__ is a very important preventive measure for early detection of this latter cancer type. The cessation of ovulation in an aging woman is called __(14)__ . Intense vasodilation of blood vessels in the skin lead to uncomfortable __(15)__ . Additionally, bone mass __(16)__ and blood levels of cholesterol __(17)__ when levels of the hormone __(18)__ wane. In contrast, healthy men are able to father children well into their eighth decade of life. Postmenopausal women are particularly susceptible to __(19)__ inflammations. The single most common problem of elderly men involves the enlargement of the __(20)__ , which interferes with the functioning of both the __(21)__ and __(22)__ systems.

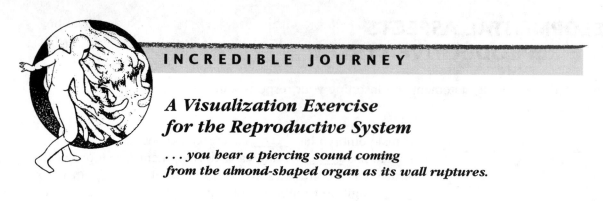

**INCREDIBLE JOURNEY**

## A Visualization Exercise for the Reproductive System

*. . . you hear a piercing sound coming from the almond-shaped organ as its wall ruptures.*

**35.** Where necessary, complete statements by inserting the missing word(s) in the answer blanks.

_____ 1.

_____ 2.

_____ 3.

_____ 4.

_____ 5.

_____ 6.

_____ 7.

_____ 8.

This is your final journey. You are introduced to a hostess this time, who has agreed to have her cycles speeded up by megahormone therapy so that all of your observations can be completed in less than a day. Your instructions are to observe and document as many events of the two female cycles as possible.

You are miniaturized to enter your hostess through a tiny incision in her abdominal wall (this procedure is called a laparotomy, or, more commonly, "belly button surgery") and end up in her peritoneal cavity. You land on a large and pear-shaped organ in the abdominal cavity midline, the __(1)__. You survey the surroundings and begin to make organ identifications and notes of your observations. Laterally and way above you on each side is an almond-shaped __(2)__, which is suspended by a ligament and almost touched by "feather-duster-like" projections of a tube snaking across the abdominal cavity toward the almond-shaped organs. The projections appear to be almost still, which is puzzling because you thought that they were the __(3)__, or fingerlike projections of the uterine tubes, which are supposed to be in motion. You walk toward the end of one of the uterine tubes to take a better look. As you study the ends of the uterine tube more closely, you discover that the featherlike projections are now moving more rapidly, as if they are trying to coax something into the uterine tube. Then you spot a reddened area on the almond-shaped organ, which seems to be enlarging even as you watch. As you continue to observe the area, you waft gently up and down in the peritoneal fluid. Suddenly you feel a gentle but insistent sucking current, drawing you slowly toward the uterine tube. You look upward and see that the reddened area now looks like an angry boil, and the uterine tube projections are gyrating and waving frantically. You realize that you are about to witness __(4)__. You try to get still closer to the opening of the uterine tube, when you hear a piercing sound coming from the almond-shaped organ as its wall ruptures. Then you see a ball-like structure, with a "halo" of tiny cells enclosing it, being drawn into the uterine tube. You have just seen the __(5)__, surrounded by its capsule of __(6)__ cells, entering the uterine tube. You hurry into the uterine tube behind it and, holding onto one of the tiny cells, follow it to the uterus. The cell mass that you have attached to has no way of propelling itself, yet you are being squeezed along toward the uterus by a process called __(7)__. You also notice that there are __(8)__, or tiny hair-like projections of the tubule cells, that are all waving in the same direction as you are moving.

_____ 9.

_____ 10.

_____ 11.

_____ 12.

_____ 13.

_____ 14.

_____ 15.

_____ 16.

_____ 17.

_____ 18.

_____ 19.

_____ 20.

Nothing seems to change as you are carried along until finally you are startled by a deafening noise. Suddenly there are thousands of tadpole-like __(9)__ swarming all around you and the sphere of cells. Their heads seem to explode as their __(10)__ break and liberate digestive enzymes. The cell mass now has hundreds of openings in it, and some of the small cells are beginning to fall away. As you peer through the rather transparent cell "halo" you see that one of the tadpole-like structures has penetrated the large central cell. Chromosomes then appear, and that cell begins to divide. You have just witnessed the second __(11)__ division. The products of this division are one large cell, the __(12)__, and one very tiny cell, a __(13)__, which is now being ejected. This cell will soon be __(14)__ because it has essentially no cytoplasm or food reserves. As you continue to watch, the sperm nucleus and that of the large central cell fuse, an event called __(15)__. You note that the new cell just formed by this fusion is called a __(16)__, the first cell of the embryonic body.

As you continue to move along the uterine tube, the central cell divides so fast that no cell growth occurs between the divisions. Thus the number of cells forming the embryonic body increases, but the cells become smaller and smaller. This embryonic division process is called __(17)__.

Finally, the uterine chamber looms before you. As you drift into its cavity, you scrutinize its lining, the __(18)__. You notice that it is thick and velvety in appearance and that the fluids you are drifting in are slightly sweet. The embryo makes its first contact with the lining, detaches, and then makes a second contact at a slightly more inferior location. This time it sticks, and as you watch, the lining of the organ begins to erode away. The embryo is obviously beginning to burrow into the rich cushiony lining, and you realize that __(19)__ is occurring.

You now leave the embryo and propel yourself well away from it. As you float in the cavity fluids, you watch the embryo disappear from sight beneath the lining. Then you continue to travel downward through your hostess's reproductive tract, exiting her body at the external opening of the __(20)__.

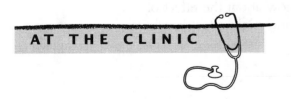

**AT THE CLINIC**

**36.** A 28-year-old primagravida (in first pregnancy) has been in the first stage of labor for several hours. Her uterine contractions are weak, and her labor is not progressing normally. Since the woman insists upon a vaginal delivery, the physician orders that Pitocin (a synthetic oxytocin) be infused. What will the effect of Pitocin be? What is the normal mechanism by which oxytocin acts to promote birth?

**37.** A 38-year-old male is upset about his low sperm count and visits a "practitioner" who commonly advertises his miracle cures for sterility. In fact, the practitioner is a quack who treats conditions of low sperm count with mega-doses of testosterone. Although his patients experience a huge surge in libido, their sperm count is even lower after hormone treatment. Explain why.

**38.** Mr. and Mrs. John Cary, a young couple who had been trying unsuccessfully to have a family for years, underwent a series of tests with a fertility clinic to try to determine the problem. Mr. Cary was found to have a normal sperm count and the sperm morphology and motility are normal.

Mrs. Cary's history sheet revealed that she had two episodes of PID during her early 20s, and the time span between successive menses ranged from 21 to 30 days. She claimed that her family was "badgering" her about not giving them grandchildren and that she was frequently despondent. A battery of hormonal tests was ordered, and Mrs. Cary was asked to perform cervical mucus testing and daily basal temperature recordings. Additionally, gas was blown through her uterine tubes to determine their patency. Her tubes proved to be closed and she was determined to be anovulatory. What do you suggest might have caused the closing of her tubes? Which of the tests done or ordered would have revealed her anovulatory condition?

**39.** A man swam in a cold lake for an hour and then noticed that his scrotum was shrunken and wrinkled. His first thought was that he had lost his testicles. What had really happened?

**40.** Mary is a heavy smoker and has ignored a friend's advice to stop smoking during her pregnancy. On the basis of what you know about the effect of smoking on physiology, describe how Mary's smoking might affect her fetus.

**41.** Mrs. Ginko's Pap smear shows some abnormal cells. What possibility should be investigated?

# *Answers*

## Chapter 1   The Human Body: An Orientation

### An Overview of Anatomy and Physiology

**1.** 1. D or physiology.    2. A or anatomy.    3. B or homeostasis.    4. C or metabolism.

**2.** Physiological study: C, D, E, F, G, H, J, K.    Anatomical study: A, B, I, K, L, M.

### Levels of Structural Organization

**3.** Cells, tissues, organs, organ systems.

**4.** 1. Atom.    2. Epithelium.    3. Heart.    4. Digestive system.

**5.** 1. K or urinary.    2. C or endocrine.    3. J or skeletal.    4. A or cardiovascular.    5. D or integumentary.
    6. E or lymphatic/immune.    7. B or digestive.    8. I or respiratory.    9. A or cardiovascular.    10. F or muscular.
    11. K or urinary.    12. H or reproductive.    13. C or endocrine.    14. D or integumentary.

**6.** 1. A or cardiovascular.    2. C or endocrine.    3. K or urinary.    4. H or reproductive.    5. B or digestive.
    6. J or skeletal.    7. G or nervous.

**7.**

**Figure 1–1:** Cardiovascular system

**Figure 1–2:** Respiratory system

**Figure 1–3:** Nervous system

**Figure 1–4:** Urinary system

**Figure 1–5:** Digestive system

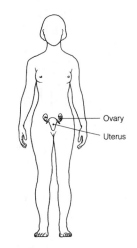

**Figure 1–6:** Reproductive system

## Maintaining Life

**8.** 1. D or maintenance of boundaries.    2. H or reproduction.    3. C or growth.    4. A or digestion.    5. B or excretion.
6. G or responsiveness.    7. F or movement.    8. E or metabolism.    9. D or maintenance of boundaries.

**9.** 1. C or nutrients.    2. B or atmospheric pressure.    3. E or water.    4. D or oxygen.    5. E or water.
6. A or appropriate body temperature.

## Homeostasis

**10.** 1. Receptor.    2. Control center.    3. Afferent.    4. Control center.    5. Effector.    6. Efferent.    7. Negative.
8. Positive.    9. Negative.

## The Language of Anatomy

**11.** 1. Ventral.    2. Dorsal.    3. Dorsal.

**12.** 1. Distal.    2. Antecubital.    3. Brachial.    4. Left upper quadrant.

**13. Figure 1–7:**

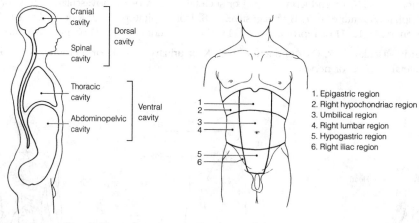

1. Epigastric region
2. Right hypochondriac region
3. Umbilical region
4. Right lumbar region
5. Hypogastric region
6. Right iliac region

**14.** 1. C or axillary.    2. G or femoral.    3. H or gluteal.    4. F or cervical.    5. P or umbilical.    6. M or pubic.
7. B or antecubital.    8. K or occipital.    9. I or inguinal.    10. J or lumbar.    11. E or buccal.

**15. Figure 1–8:** Section A: Midsagittal.    Section B: Transverse.

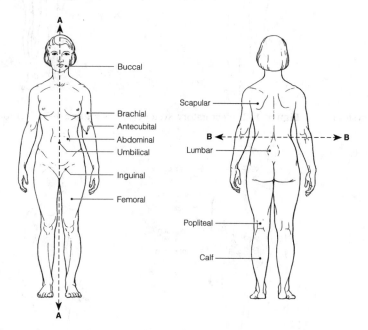

**16.** 1. G or ventral, D or pelvic.    2. G or ventral, F or thoracic.    3. C or dorsal, B or cranial.    4. G or ventral,
D or pelvic.    5. G or ventral, A or abdominal.

**17.** 1. A or anterior.    2. G or posterior.    3. J or superior.    4. J or superior.    5. E or lateral.    6. A or anterior.
7. F or medial.    8. H or proximal.    9. B or distal.    10. G or posterior.    11. J or superior.    12. I or sagittal.
13. C or frontal.    14. C or frontal.    15. K or transverse.

**18.** 1–5. A or abdominopelvic.    6. C or spinal.    7. A or abdominopelvic.    8 and 9. D or thoracic.    10. B or cranial.
11 and 12. A or abdominopelvic.

**19.** 1. 2, 3, 7, 11, 12.   2. 2, 3.   3. 2.   4. 1, 2, 3, 5.   5. 2, 3.

**At the Clinic**

**20.** Skeletal, muscular, cardiovascular, integumentary, nervous.

**21.** The need for nutrients and water.

**22.** The anterior and lateral aspects of the abdomen have no bony (skeletal) protection.

**23.** John has a hernia in the area where his thigh and trunk meet, pain from his infected kidney radiating to his lower back, and bruises in his genital area.

# Chapter 2   Basic Chemistry

## Concepts of Matter and Energy

**1.** 1. B, D.   2. A, B, C, D.   3. A, B.

**2.** 1. C or mechanical.   2. B, D or electrical, radiant.   3. C or mechanical.   4. A or chemical.   5. D or radiant.

## Composition of Matter

**3.**

| Particle | Location | Electrical charge | Mass |
|----------|----------|-------------------|------|
| Proton | Nucleus | +1 | 1 amu |
| Neutron | Nucleus | 0 | 1 amu |
| Electron | Orbitals | −1 | 0 amu |

**4.** 1. O.   2. C.   3. K.   4. I.   5. H.   6. N.   7. Ca.   8. Na.   9. P.   10. Mg.   11. Cl.   12. Fe.

**5.** 1. E or ion.   2. F or matter.   3. C or element.   4. B or electrons.   5. B or electrons.   6. D or energy.   7. A or atom.   8. G or molecule.   9. I or protons.   10. J or valence.   11 and 12. H and I or neutrons and protons.

**6.** 1. *T.*   2. Protons.   3. More.   4. *T.*   5. Radioactive.   6. *T.*   7. Chloride.   8. Iodine.   9. *T.*

### Molecules, Chemical Bonds, and Chemical Reactions

**7.** 1. C or synthesis.   2. B or exchange.   3. A or decomposition.

**8.** **Figure 2–1:** The nucleus is the innermost circle containing 6P and 6N; the electrons are indicated by the small circles in the orbits.   1. Atomic number is 6.   2. Atomic mass is 12 amu.   3. Carbon.   4. Isotope.   5. Chemically active.   6. Four electrons.   7. Covalent because it would be very difficult to gain or lose four electrons.

**9.** $H_2O_2$ is one molecule of hydrogen peroxide (a compound). $2OH^-$ represents two hydroxide ions.

**10.** **Figure 2–2: A** represents an ionic bond; **B** shows a covalent bond.

**11. Figure 2–3**

**12.** Circle B, C, E.

**13.** 1. $H_2CO_3$.   2. $H^+$ and $HCO_3^-$.   3. The ions should be circled.   4. An additional arrow going to the left should be added between $H_2CO_3$ and $H^+$.

## Biochemistry: The Composition of Living Matter

**14.** 1–3. A or acid(s), B or base(s), and D or salt(s).   4. B or base(s).   5. A or acid(s).   6. D or salt(s).
7. D or salt(s).   8. A or acid(s).   9. C or buffer.

**15.** 1. Heat capacity.   2. Water.   3. 70% (60-80%).   4. Hydrogen.   5 and 6. Hydrolysis and dehydration.
7. Polarity.   8. Lubricants.

**16.** Weak acid: B, C, E. Strong acid: A, D, E, F, G.

**17.** *X* fats, proteins, glucose, and DNA.

**18.** 1. G or monosaccharides.   2. D or fatty acids and E or glycerol.   3. A or amino acids.   4. F or nucleotides.
5. H or proteins.   6. G or monosaccharides (B or carbohydrates).   7. C or lipids.   8. G or monosaccharides (B or carbohydrates).   9. C or lipids.   10 and 11. F or nucleotides and A or amino acids.   12. F or nucleotides.
13. C or lipids.   14. H or proteins.   15. B or carbohydrates.   16. C or lipids.   17. H or proteins.   18. C or lipids.

**19.** 1. B or collagen, H or keratin.   2. D or enzyme, F or hemoglobin, some of G or hormones.   3. D or enzyme.
4. L or starch.   5. E or glycogen.   6. C or DNA.   7. A or cholesterol (Some G or hormones are steroids).
8. I or lactose, J or maltose.

**20. Figure 2–4:** A. Monosaccharide.   B. Globular protein.   C. Nucleotide.   D. Fat.   E. Polysaccharide.

**21.** 1. Glucose.   2. Ribose.   3. Glycogen.   4. Glycerol.   5. Glucose.

**22.** 1. *T.*   2. Neutral fats.   3. *T.*   4. Polar.   5. *T.*   6. ATP.   7. *T.*   8. O.

**23.** Unnamed nitrogen bases: Thymine (T) and Guanine (G). 1. Hydrogen bonds.   2. Double helix.   3. 12.
4. Complementary.

**Figure 2–5:**

DNA

Note that the stippled parts of the backbones represent phosphate units (P) while the unaltered (white) parts of the backbones that are attached to the bases are deoxyribose sugar (d-R) units.

**24.** 1. C or glucose.   2. C or enter between the monomers, etc.   3. B or hydrolysis.   4. A or R group.

## The Incredible Journey

**25.** 1. Negative.   2. Positive.   3. Hydrogen bonds.   4. Red blood cells.   5. Protein.   6. Amino acids.   7. Peptide.
8. $H^+$ and $OH^-$.   9. Hydrolysis.   10. Enzyme.   11. Glucose.   12. Glycogen.   13. Dehydration synthesis.
14. $H_2O$.   15. Increase.

**At the Clinic**

26. Acidosis means blood pH is below the normal range. The patient should be treated with something to raise the pH.

27. Each of the 20 amino acids has a different chemical group called the R group. The R group on each amino acid determines how it will fit in the folded, three-dimensional, tertiary structure of the protein and the bonds it may form. If the wrong amino acid is inserted, its R group might not fit into the tertiary structure properly, or required bonds might not be made; hence the entire structure might be altered. Because function depends on structure, this means the protein will not function properly.

28. Heat increases the kinetic energy of molecules. Vital biological molecules, like proteins and nucleic acids, are denatured (rendered nonfunctional) by excessive heat because intramolecular bonds essential to their functional structure are broken. Because all enzymes are proteins, their destruction is lethal.

29. An MRI because it allows visualization of soft structures enclosed by bone (e.g., the skull).

# Chapter 3   Cells and Tissues

**Cells**

1. 1–4. (in any order): Carbon; Oxygen; Nitrogen; Hydrogen.   5. Water.   6. Calcium.   7. Iron.   8–12. (five of the following, in any order): Metabolism; Reproduction; Irritability; Mobility; Ability to grow; Ability to digest foods; Ability to excrete waste.   13–15. (three of the following, in any order): Cubelike; Tilelike; Disk-shaped; Round spheres; Branching; Cylindrical.   16. Interstitial (tissue) fluid.   17. Squamous epithelial.

2.

| Cell structure | Location | Function |
|---|---|---|
| Plasma membrane | External boundary of the cell | Confines cell contents; regulates entry and exit of materials |
| Lysosomes | Scattered in cytoplasm | Digest ingested materials and worn-out organelles |
| Mitochondria | Scattered throughout the cell | Control release of energy from foods; form ATP |
| Microvilli | Projections of the plasma membrane | Increase the membrane surface area |
| Golgi apparatus | Near the nucleus (in the cytoplasm) | Packages proteins to be incorporated into the plasma membrane or lysosomes or exported from the cell |
| Nucleus | (Usually) center of the cell | Storehouse of genetic information; directs cellular activities, including division |
| Centrioles | Two rod-shaped bodies near the nucleus | Direct formation of the mitotic spindle |
| Nucleolus | Dark spherical body in the nucleus | Storehouse/assembly site for ribosomes |
| Smooth ER | In the cytoplasm | Site of steroid synthesis and lipid metabolism |
| Rough ER | In the cytoplasm | Transports proteins (made on its ribosomes) to other sites in the cell; synthesizes membrane lipids |
| Ribosomes | Attached to membrane systems or scattered in the cytoplasm | Synthesize proteins |
| Chromatin | Dispersed in the nucleus | Contains genetic material (DNA); coils during mitosis |
| Peroxisomes | Scattered in the cytoplasm | Detoxify alcohol; hydrogen peroxide, etc. |
| Inclusions | Dispersed in the cytoplasm | Provide nutrients; represent cell waste, stored products, etc. |

3. **Figure 3–1:**

Smooth ER

Cytosol
Vacuole

Centriole(s)

Microvilli

Mitochondrion

Nuclear membrane

Golgi apparatus
Nucleolus
Chromatin
thread

Rough ER

Plasma membrane

4. 1. A.   2. B.   3. C.   4. A.

5. **Figure 3–2:** 1. A; Crenated.   2. B; The same solute concentration inside and outside the cell.   3. C; They are bursting (lysis); water is moving by osmosis from its site of higher concentration (cell exterior) into the cell where it is in lower concentration, causing the cells to swell.

6. 1. C or endocytosis, phagocytosis, D or exocytosis, F or solute pumping.   2. A or diffusion, simple, B or diffusion, osmosis.   3. E or filtration.   4. A or diffusion, simple, B or diffusion, osmosis.   5. F or solute pumping. 6. D or exocytosis.   7. B or diffusion, osmosis.   8. F or solute pumping.   9. C or endocytosis, phagocytosis. 10. A or diffusion, simple.

7. **Figure 3–3:** Oxygen and fats will move from the cell exterior into the cell by moving passively through the lipid portion; the lipid portion is indicated as the major part of the membrane composed of small spheres, each with two "tails." Amino acids and glucose also enter the cell from the exterior but do so by attaching to a protein carrier (indicated as large, solid, irregularly shaped structures extending part or all of the way though the membrane). Carbon dioxide, like oxygen, passes by diffusion through the lipid part of the membrane, but in the opposite direction (i.e., from the cell interior to the cell exterior).

Cholesterol and sugar groups (which attach to externally facing proteins) are not shown in the diagram.

8. 1. P or proteins.   2. K or helix.   3. O or phosphate.   4. T or sugar.   5. C or bases.   6. B or amino acids. 7. E or complementary.   8. F or cytosine.   9. V or thymine.   10. S or ribosome.   11. Q or replication. 12. M or nucleotides.   13. U or template, or model.   14. L or new.   15. N or old.   16. H or genes. 17. I or growth.   18. R or repair.

9. **Figure 3–4:** A. Prophase.   B. Anaphase.   C. Telophase.   D. Metaphase.

Nuclear
membrane

Chromosomes

Spindle fibers

Centrioles

Centrioles

Chromosomes

Nucleoli

Nuclear
membrane

Spindle fibers

10. 1. C or prophase.   2. A or anaphase.   3. D or telophase.   4. D or telophase.   5. B or metaphase. 6. C or prophase.   7. C or prophase.   8. E or none of these.   9. C or prophase.   10. C or prophase. 11. D or telophase.   12. A or anaphase, B or metaphase.   13. E or none of these.

11. 1. Nucleus.   2. Cytoplasm.   3. Coiled.   4. Centromeres.   5. Binucleate cell.   6. Spindle.   7. Interphase.

**12. Figure 3–5:** 1. transcription.    2. translation.    3. anticodon; triplet.

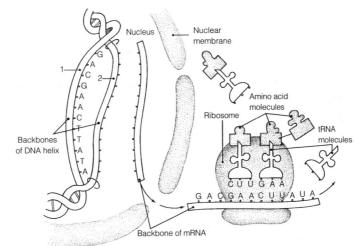

**Body Tissues**

**13. Figure 3–6:** A. Simple squamous epithelium.    B. Simple cuboidal epithelium.    C. Cardiac muscle.
D. Dense fibrous connective tissue.    E. Bone.    F. Skeletal muscle.    G. Nervous tissue.
H. Hyaline cartilage.    I. Smooth muscle tissue.    J. Adipose (fat) tissue.    K. Stratified squamous epithelium.
L. Areolar connective tissue. The noncellular portions of D, E, H, J, and L are matrix.

**14.** The neuron has long cytoplasmic extensions that promote its ability to transmit impulses over long distances
within the body.

**15.** 1. B or epithelium.    2. C or muscle.    3. D or nervous.    4. A or connective.    5. B or epithelium.
6. D or nervous.    7. C or muscle.    8. B or epithelium.    9. A or connective.    10. A or connective.
11. C or muscle.    12. A or connective.    13. D or nervous.

**16.** 1. E or stratified squamous.    2. B or simple columnar.    3. E or stratified squamous.    4. A or pseudostratified
columnar (ciliated).    5. A or pseudostratified columnar (ciliated).    6. F or transitional.    7. D or simple squamous.

**17. Figure 3–7:**

**18.** 1. Skeletal.    2. Cardiac; smooth.    3. Skeletal; cardiac.    4. Smooth (most cardiac).    5–7. Skeletal.
8 and 9. Smooth.    10. Cardiac.    11. Skeletal.    12. Cardiac.    13. Skeletal.    14. Smooth, cardiac.    15. Cardiac.

**19.** 1. Cell.    2. Elastic fibers.    3. Bones.    4. Nervous.    5. Blood.

**20.** 1. C or dense fibrous.    2. A or adipose.    3. C or dense fibrous.    4. D or osseous tissue.    5. B or areolar.
6. F or hyaline cartilage.    7. A or adipose.    8. F or hyaline cartilage.    9. D or osseous tissue.    10. E or reticular.

**21.** 1. Inflammation.    2. Clotting proteins.    3. Granulation.    4. Regeneration.    5. *T.*    6. Collagen.    7. *T.*

**Developmental Aspects of Cells and Tissues**

**22.** 1. Tissues.    2. Growth.    3. Nervous.    4. Muscle.    5. Connective (scar).    6. Chemical.    7. Physical.
8. Genes (DNA).    9–11. Connective tissue changes; Decreased endocrine system activity; Dehydration of body tissues.
12. Division.    13 and 14. Benign; Malignant.    15. Benign.    16. Malignant.    17. Biopsy.    18. Surgical removal.
19. Hyperplasia.    20. Atrophy.

**The Incredible Journey**

**23.** 1. Cytoplasm (cytosol).    2. Nucleus.    3. Mitochondrion.    4. ATP.    5. Ribosomes.    6. Rough endoplasmic
reticulum.    7. Pores.    8. Chromatin.    9. DNA.    10. Nucleoli.    11. Golgi apparatus.    12. Lysosome.

**At the Clinic**

**24.** The oxidases of ruptured peroxisomes were converting the hydrogen peroxide to water and (free) oxygen gas (which causes the bubbling).

**25.** Generally speaking, stratified epithelia consisting of several cell layers are more effective where abrasion is a problem than are simple epithelia (consisting of one cell layer).

**26.** Streptomycin inhibits bacterial protein synthesis. If the bacteria are unable to synthesize new proteins (many of which would be essential enzymes), they will die.

**27.** Since connective tissue is the most widespread tissue in the body, and is found either as part of or is associated with every body organ, the physician will most likely tell her that she can expect the effects of lupus to be very diffuse and widespread.

**28.** Granulation tissue secretes substances that kill bacteria.

# Chapter 4    Skin and Body Membranes

## Classification of Body Membranes

**1.** The mucous, serous, and cutaneous membranes are all composite membranes composed of an epithelial layer underlaid by a connective tissue layer. A mucous membrane is an epithelial sheet underlaid by a connective tissue layer called the lamina propria. Mucosae line the respiratory, digestive, urinary, and reproductive tracts; functions include protection, lubrication, secretion, and absorption. Serous membranes consist of a layer of simple squamous epithelium resting on a scant layer of fine connective tissue. Serosae line internal ventral body cavities and cover their organs; their function is to produce a lubricating fluid that reduces friction. The cutaneous membrane, or skin, is composed of the epithelial epidermis and the connective tissue dermis. It covers the body exterior and protects deeper body tissues from external insults. The synovial membranes, which line joint cavities of synovial joints, are composed entirely of connective tissue. They function to produce lubrication to decrease friction within the joint cavity.

**2.** In each case, the visceral layer of the serosa covers the external surface of the organ, and the parietal layer lines the body cavity walls.

**Figure 4–1:**

## Integumentary System (Skin)

**3. Figure 4–2:**

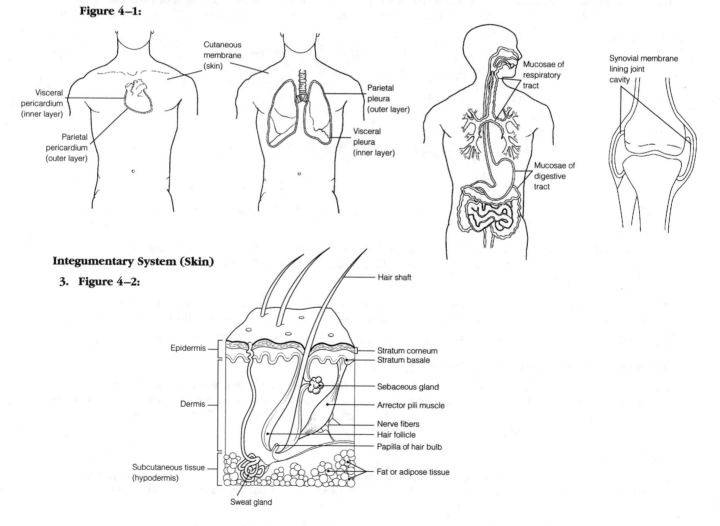

**4.** 1. As the basal cells continue to divide, the more superficial cells are pushed farther and farther from the nutrient supply diffusing from the dermis.   2. Water-proofing substances (keratin and others) made by the keratinocytes effectively limit nutrient entry into the cells.

**5.** 1. Heat.   2. Subcutaneous.   3. Keratin.   4. Vitamin D.   5. A freckle.   6. Elasticity.   7. Oxygen (blood flow).   8. Cyanosis.

**6.** 1. A or stratum corneum, D or stratum lucidum.   2. A or stratum corneum, D or stratum lucidum.   3. E or papillary layer.   4. H or dermis as a whole.   5. B or stratum basale.   6. A or stratum corneum.   7. H or dermis as a whole.   8. B or stratum basale.   9. G or epidermis as a whole.

**7.** 1. Melanin.   2. Keratin.   3. *T.*   4. Stratum corneum.   5. Shaft.   6. Dermis.

**8. Figure 4–3:**

**9.** 1. E or sebaceous glands.   2. A or arrector pili.   3. G or eccrine sweat glands.   4. D or hair follicle.   5. F or apocrine sweat gland.   6. C or hair.   7. B or cutaneous receptors.   8. E or sebaceous glands, and F or apocrine sweat glands.   9. G or eccrine sweat glands.

**10.** 1. Arrector pili.   2. Absorption.   3. Stratum basale.   4. Blackheads.   5. Eccrine glands.   6. Wrinkles.   7. Keratin.

**11.** Water/protein/electrolyte loss; circulatory collapse, renal shutdown.

**12.** 1. C or third-degree burn.   2. B or second-degree burn.   3. A or first-degree burn.   4. B or second-degree burn.   5. C or third-degree burn.   6. C or third-degree burn.

**13.** It allows estimation of the extent of burns so that fluid volume replacement can be correctly calculated.

**14.** 1. Squamous cell carcinoma.   2. Basal cell carcinoma.   3. Malignant melanoma.

**15.** Pigmented areas that are <u>A</u>ssymetric, have irregular <u>B</u>orders, exhibit several <u>C</u>olors, and have a <u>D</u>iameter greater than 6 mm are likely to be cancerous.

### Developmental Aspects of the Skin and Body Membranes

**16.** 1. C or dermatitis.   2. D or delayed-action gene.   3. F or milia.   4. B or cold intolerance.   5. A or acne.   6. G or vernix caseosa.   7. E or lanugo.

### The Incredible Journey

**17.** 1. Collagen.   2. Elastin (or elastic).   3. Dermis.   4. Phagocyte (macrophage).   5. Hair follicle connective tissue.   6. Epidermis.   7. Stratum basale.   8. Melanin.   9. Keratin.   10. Squamous (stratum corneum) cells.

### At the Clinic

**18.** Chemotherapy drugs used to treat cancer kill the most rapidly dividing cells in the body, including many matrix cells in the hair follicles; thus, the hair falls out.

**19.** The baby has seborrhea, or cradle cap, a condition of overactive sebaceous glands. It is not serious; the oily deposit is easily removed with attentive washing, and soon stops forming.

**20.** Bedridden patients are turned at regular intervals so that no region of their body is pressed against the bed long enough to deprive the blood supply to that skin; thus, bedsores are avoided.

**21.** Porphyria (a) may have been Count Dracula's disease.

**22.** Besides storing fat as a source of nutrition, the hypodermis anchors the skin to underlying structures (such as muscles) and acts as an insulator against heat loss.

**23.** The body of a nail is its visible, attached part (not its white free edge). The root is the proximal part that is embedded in skin. The bed is the part of the epidermis upon which the nail lies. The matrix is the proximal part of the nail bed and it is responsible for nail growth. The cuticle is the skin fold around the perimeter of the nail body. Since the matrix is gone, the nail will not grow back.

**24.** The peritoneum will be inflamed and infected. Since the peritoneum encloses so many richly vascularized organs, a spreading peritoneal infection can be life threatening.

## Chapter 5   The Skeletal System

### Bones—An Overview

**1.** 1. P.   2. P.   3. D.   4. D.   5. P.   6. D.   7. P.   8. P.   9. P.

**2.** 1. S.   2. F.   3. L.   4. L.   5. F.   6. L.   7. L.   8. F.   9. I.

**3.** 1. C or epiphysis.   2. A or diaphysis.   3. C or epiphysis, D or red marrow.   4. A or diaphysis.   5. E or yellow marrow cavity.   6. B or epiphyseal plate.

**4.** 1. G or parathyroid hormone.   2. F or osteocytes.   3. A or atrophy.   4. H or stress/tension.   5. D or osteoblasts.   6. B or calcitonin.   7. E or osteoclasts.   8. C or gravity.

**5. Figure 5–1:** 1. B or concentric lamellae.   2. C or lacunae.   3. A or central (Haversian) canal.   4. E or bone matrix.   5. D or canaliculi.

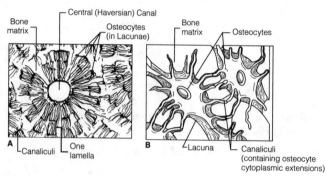

**6.** 1. Yellow marrow.   2. Osteoblasts.   3. Marrow cavity.   4. Periosteum.

**7. Figure 5–2:** The epiphyseal plate is the white band shown in the center region of the head; the articular cartilage is the white band on the external surface of the head. Red marrow is found within the spongy bone cavities; yellow marrow is found within the cavity of the diaphysis.

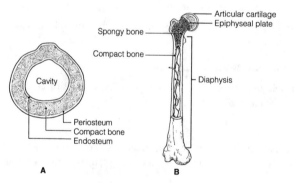

### Axial Skeleton

**8.** 1. B or frontal.   2. N or zygomatic.   3. E or mandible.   4. G or nasals.   5. I or palatines.   6. J or parietals.   7. H or occipital.   8. K or sphenoid.   9. D or lacrimals.   10. F or maxillae.   11. A or ethmoid.   12. L or temporals.   13. K or sphenoid.   14. A or ethmoid.   15. E or mandible.   16. L or temporals.   17–20. A or ethmoid, B or frontal, F or maxillae, and K or sphenoid.   21. H or occipital.   22. H or occipital.   23. L or temporals.   24. M or vomer.   25. A or ethmoid.

9. **Figure 5–3:**

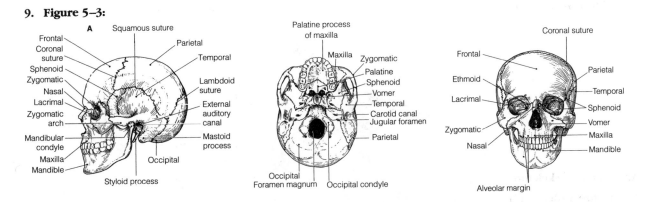

10. **Figure 5–4:** 1. Mucosa-lined air-filled cavities in bone.   2. They lighten the skull and serve as resonance chambers for speech.   3. Their mucosa is continuous with that of the nasal passages into which they drain.

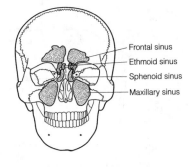

11. 1. F or vertebral arch.   2. A or body.   3. C or spinous process, E or transverse process.   4. A or body, E or transverse process.   5. B or intervertebral foramina.

12. 1. A or atlas, B or axis, C or cervical vertebra—typical.   2. B or axis.   3. G or thoracic vertebra.   4. F or sacrum.   5. E or lumbar vertebra.   6. D or coccyx.   7. A or atlas.   8. A or atlas, B or axis, and C or cervical vertebra—typical.   9. G or thoracic vertebra.

13. 1. Kyphosis.   2. Scoliosis.   3. Fibrocartilage.   4. Springiness or flexibility.

14. **Figure 5–5:** A. Cervical; atlas.   B. Cervical.   C. Thoracic.   D. Lumbar.

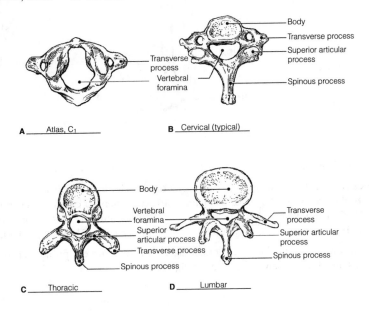

15. **Figure 5–6:** 1. Cervical, $C_1$–$C_7$.   2. Thoracic, $T_1$–$T_{12}$.   3. Lumbar, $L_1$–$L_5$.   4. Sacrum, fused.   5. Coccyx, fused.   6. Atlas, $C_1$.   7. Axis, $C_2$.

16. 1. Lungs.   2. Heart.   3. True.   4. False.   5. Floating.   6. Thoracic vertebrae.   7. Sternum.   8. An inverted cone.

**17. Figure 5–7:** Ribs #1–#7 on each side are true ribs; ribs #8–#12 on each side are false ribs.

**Appendicular Skeleton**

**18. Figure 5–8:** Scapula.

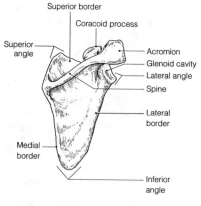

**19. Figure 5–9:** A. Humerus.    B. Ulna.    C. Radius.        **20. Figure 5–10:**

**21.** Pectoral: A, C, D. Pelvic: B, E, F.

**22.** 1. G or deltoid tuberosity.    2. I or humerus.    3 and 4. D or clavicle, P or scapula.    5 and 6. O or radius, T or ulna.
7. A or acromion.    8. P or scapula.    9. D or clavicle.    10. H or glenoid cavity.    11. E or coracoid process.
12. D or clavicle.    13. S or trochlea.    14. T or ulna.    15. B or capitulum.    16. F or coronoid fossa.
17. T or ulna.    18 and 19. P or scapula, Q or sternum.    20. C or carpals.    21. M or phalanges.    22. J or metacarpals.

**23.** 1. Female inlet is larger and more circular.    2. Female sacrum is less curved; pubic arch is rounder.
3. Female ischial spines are shorter; pelvis is shallower/lighter.

**Figure 5–11:**

**24.** 1. Ulna.    2. Pelvis.    3. Scapula.    4. Mandible.    5. Carpals.

**25.** 1. I or ilium, K or ischium, S or pubis.    2. J or ischial tuberosity.    3. R or pubic symphysis.    4. H or iliac crest.
5. A or acetabulum.    6. T or sacroiliac joint.    7. C or femur.    8. D or fibula.    9. W or tibia.    10. C or femur,
Q or patella, W or tibia.    11. X or tibial tuberosity.    12. Q or patella.    13. W or tibia.    14. N or medial
malleolus.    15. L or lateral malleolus.    16. B or calcaneus.    17. V or tarsals.    18. O or metatarsals.    19. P or
obturator foramen.    20. G or greater and lesser trochanters, E or gluteal tuberosity.    21. U or talus.

**26.** 1. Pelvic.    2. Phalanges.    3. *T.*    4. Acetabulum.    5. Sciatic.    6. *T.*    7. Hip bones.    8. *T.*

**27. Figure 5–12:**

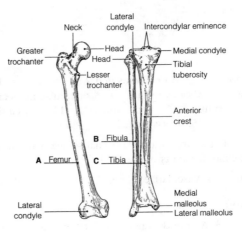

**28. Figure 5–13:** Bones of the skull, vertebral column, and bony thorax are parts of the axial skeleton. All others belong
to the appendicular skeleton.

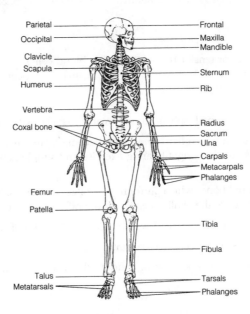

**Bone Fractures**

**29.**  1. G or simple fracture.    2. A or closed reduction.    3. E or greenstick fracture.    4. B or compression fracture.    5. C or compound fracture.    6. F or open reduction.    7. H or spiral fracture.

**Figure 5–14:**

**30.**  1. *T.*    2. *T.*    3. Phagocytes (macrophages).    4. *T.*    5. Periosteum.    6. *T.*    7. Spongy.

**Joints**

**31.**  1. Synovial fluid.    2. Articular cartilage.    3. Ligaments.

**Figure 5–15:**

**32.**  1. A or cartilaginous.    2. C or synovial.    3. B or fibrous and 2 or suture.    4. B or fibrous and 2 or suture.    5. C or synovial.    6. C or synovial.    7. C or synovial.    8. A or cartilaginous and 3 or symphysis.    9. C or synovial.    10. B or fibrous and 2 or suture.    11. C or synovial.    12. A or cartilaginous and 1 or epiphyseal disk.    13. C or synovial.    14. C or synovial.    15. C or synovial.

**33.**  Synovial joints are diarthroses or freely movable joints. The axial skeleton supports and protects internal organs; thus, strength is more important than mobility for joints of the axial skeleton.

**34.**  1. *T.*    2. Osteoarthritis.    3. Acute.    4. Vascularized.    5. *T.*    6. Gouty arthritis or gout.    7. Rickets.    8. *T.*

**Developmental Aspects of the Skeleton**

**35.**  1. D or nervous.    2. F or urinary.    3. A or endocrine.    4. C or muscular.    5. A or endocrine.    6. B or integumentary.

**36.**  1. Fontanels.    2. Compressed.    3. Growth.    4. Sutures.    5. Thoracic.    6. Sacral.    7. Primary.    8. Cervical.    9. Lumbar.

**The Incredible Journey**

**37.**  1. Femur.    2. Spongy.    3. Stress (or tension).    4. RBCs (red blood cells).    5. Red marrow.    6. Nerve.    7. Central or Haversian.    8. Compact.    9. Canaliculi.    10. Lacunae (osteocytes).    11. Matrix.    12. Osteoclast.

**At the Clinic**

**38.**  Seven bones contribute to the orbit: frontal, sphenoid, zygomatic, maxilla, palatine, lacrimal, and ethmoid bones.

**39.**  Mrs. Bruso has severe osteoporosis in which her bones have become increasingly fragile. The postmenopausal deficit of estrogen has placed her bones at risk. Weight-bearing exercise and supplemental calcium will probably be prescribed.

**40.**  The cribriform plates of the ethmoid bone, which surround the olfactory nerves. These plates are quite fragile and are often crushed by a blow to the front of the skull. This severs the olfactory nerve fibers, which cannot grow back.

**41.**  Rheumatoid arthritis, fairly common in middle-aged women, causes this type of deformity.

**42.**  James has all the classic signs and symptoms of osteoarthritis.

**43.**  Janet will be watched for signs of scoliosis because of injury to thoracic vertebrae (and probably associated muscles) on <u>one</u> side of the body.

# Chapter 6    The Muscular System

### Overview of Muscle Tissues

**1.** 1. A or cardiac, B or smooth.    2. A or cardiac, C or skeletal.    3. B or smooth.    4. C or skeletal.    5. A or cardiac.    6. A or cardiac.    7. C or skeletal.    8. C or skeletal.    9. C or skeletal.

**2.** A. Smooth muscle.    B. Cardiac muscle.

**3.** 1. Bones.    2. Promotes labor during birth.    3. Contractility.    4. Stretchability.    5. Promotes growth.

### Microscopic Anatomy of Skeletal Muscle

**4.** 1. G or perimysium.    2. B or epimysium.    3. I or sarcomere.    4. D or fiber.    5. A or endomysium.    6. H or sarcolemma.    7. F or myofibril.    8. E or myofilament.    9. K or tendon.    10. C or fascicle.

**Figure 6–2:** The endomysium is the connective tissue that surrounds each muscle cell (fiber).

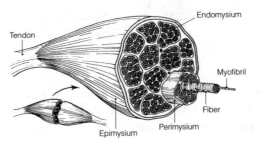

**5.** **Figure 6–3:** In the student art of a contracted sarcomere, the myosin filaments should nearly touch the Z discs and the opposing actin filaments should touch each other.

### Skeletal Muscle Activity

**6.** 1. Motor unit.    2. Axonal terminals.    3. Synaptic cleft.    4. Acetylcholine.    5. Nerve impulse (or action potential).    6. Depolarization.

**7.** **Figure 6–4:**

**8.** 1. 1.    2. 4.    3. 7.    4. 2.    5. 5.    6. 3.    7. 6.

**9.** 1. F.    2. E.    3. C.    4. B.    5. H.    6. G.    7. I.

**10.** 1. G or tetanus.    2. B or isotonic contraction.    3. I or many motor units.    4. H or few motor units.    5. A or fatigue.    6. E or isometric contraction.

**11.** 1. B.   2. C.   3. A.   4. A, B.   5–7. C.   8. B.   9. A.

**12.** Your rate of respiration (breathing) is much faster and you breathe more deeply.

**13.** Check #1, #3, #4, and #7.

## Muscle Movements, Types, and Names

**14. Figure 6–5:**

**15.** 1. Plantar flexion.   2. Dorsiflexion.   3. Circumduction.   4. Adduct.   5. Flexion.   6. Extension.   7. Extension.   8. Flexed.   9. Flexion.   10. Rotation.   11. Circumduction.   12. Rotation.   13. Pronation.   14. Abduction.

**16.** 1. C or prime mover.   2. B or fixator.   3. D or synergist.   4. D or synergist.   5. A or antagonist.   6. B or fixator.

**17.** 1. E, G.   2. A, G.   3. D, E.   4. E, F.   5. A, C, E.   6. B.   7. E, F.   8. E, F.

## Gross Anatomy of the Skeletal Muscles

**18. Figure 6–6:** 1. I.   2. A.   3. D.   4. B.   5. E.   6. C.   7. G.   8. F.

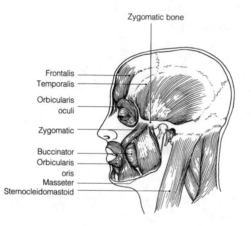

**19. Figure 6–7:** 1. I.   2. H.   3. A.   4. D.   5. J.   6. F.   7. K.   8. C.   9. B.

**20.** **Figure 6–8:** 1. F.   2. E.   3. A.   4. B.   5. E.

Trapezius

Deltoid

Latissimus dorsi

**21.** **Figure 6–9:** 1. H.   2. E.   3. D.
4. O.   5. A.   6. I.   7. G.   8. F.
9. C.   10. K.   11. N.

Gluteus medius
Gluteus maximus
Adductor muscles
Semitendinosus
Biceps femoris
Semimembranosus
Hamstring group
Sartorius
Rectus femoris
Vastus medialis
Vastus lateralis
Quadriceps group
Gastrocnemius
Fibularis muscles
Tibialis anterior
Soleus

**22.** **Figure 6–10:** 1. E.   2. D.   3. F.
4. A.   5. G.   6. B.

Deltoid
Biceps brachii
Triceps brachii
Extensor digitorum
Flexor carpi ulnaris

**23.** 1–3. Deltoid; Gluteus maximus; Gluteus medius.    4. Quadriceps.    5. Calcaneal (Achilles).    6. Proximal.
7. Forearm.    8. Anterior.    9. Posteriorly.    10. Knee.    11. Flex.

**24.** 1. Biceps femoris.    2. Antagonists.    3. Frontalis.    4. Vastus medialis.

**25.** 1. 4.    2. 5.    3. 17.    4. 16.    5. 7.    6. 6.    7. 19.    8. 14.    9. 18.    10. 12.    11. 11.    12. 10.    13. 21.
14. 1.    15. 2.    16. 3.    17. 15.    18. 20.    19. 13.    20. 9.    21. 8.

**26.** 1. 2.    2. 1.    3. 5.    4. 9.    5. 7.    6. 4.    7. 12.    8. 3.    9. 8.    10. 10.    11. 11.    12. 6.

## Developmental Aspects of the Muscular System

**27.** 1. Quickening.    2. Muscular dystrophy.    3–4. Proximal-distal and cephalocaudal.    5. Gross.    6. Fine.
7. Exercised.    8. Atrophy.    9. Myasthenia gravis.    10. Weight.    11. Size and mass.    12. Connective (scar).

## The Incredible Journey

**28.** 1. Endomysium.    2. Motor unit.    3. Neuromuscular.    4. Acetylcholine.    5. Sodium.    6. Action potential.
7. Calcium.    8. Actin.    9. Myosin.    10. Calcium.

## At the Clinic

**29.** Tendons attaching at the anterior wrist are involved in wrist and finger flexion. Malcolm will lose his ability to make a fist and grasp a baseball.

**30.** The hamstrings can be strained (pulled) when the hip is flexed and the knee is vigorously extended at the same time.

**31.** The rectus abdominis is a narrow, medially placed muscle that does not extend completely across the iliac regions. No, if the incision was made as described, the rectus abdominis was not cut.

**32.** The latissimus dorsi and the trapezius, which together cover most of the superficial surface of the back, are receiving most of the massage therapist's attention.

**33.** The chances are good that the boy has Duchenne muscular dystrophy. This condition is fatal when it impairs the respiratory muscles.

**34.** By reducing the size of the abdomen, the abdominal contents are forced into a smaller space which would increase the intra-abdominal pressure. The rise in intra-abdominal pressure would, in turn, force the vertebrae to move farther apart, reducing vertebral compression and pressure on the nerve fibers that transmit pain.

**35.** Chemical A would be a better muscle relaxant because it prevents stimulation of the muscle cells by preventing ACh from binding to the ACh receptors on the sarcolemma. Chemical B would lead to greater muscle irritability and stronger, more sustained muscle contraction, exactly the opposite of what we are seeking to accomplish.

# Chapter 7    The Nervous System

**1.** 1. It monitors all information about changes occurring both inside and outside the body.    2. It processes and interprets the information received and integrates it in order to make decisions.    3. It commands responses by activating muscles, glands, and other parts of the nervous system.

## Organization of the Nervous System

**2.** 1. B or CNS.    2. D or somatic nervous system.    3. C or PNS.    4. A or autonomic nervous system.
5. B or CNS.    6. C or PNS.

## Nervous Tissue—Structure and Function

**3.** 1. B or neuroglia.    2–4. A or neurons.    5. B or neuroglia.

**4.** 1. B or axonal terminal.    2. C or dendrite.    3. D or myelin sheath.    4. E or cell body.    5. A or axon.

**5.** 1. A or bare nerve endings, D or muscle spindle.    2. A or bare nerve endings, E or Pacinian corpuscle.
3. E or Pacinian corpuscle (perhaps also B and D).    4. B or Golgi tendon organ, D or muscle spindle.
5. C or Meissner's corpuscle.

**6.** 1. C or cutaneous sense organs.    2. L or Schwann cells.    3. M or synapse.    4. O or tract.    5. B or association neuron.    6. I or nodes of Ranvier.    7. E or ganglion.    8. D or efferent neuron.    9. K or proprioceptors.
10. N or stimuli.    11. A or afferent neuron.    12. G or neurotransmitters.

**7. Figure 7–1:**

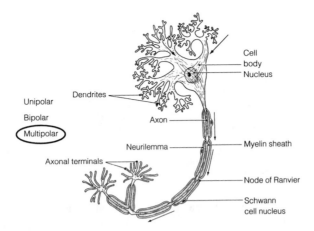

**8.** 1. Stimulus.    2. Receptor.    3. Afferent neuron.    4. Efferent neuron.    5. Effector organ.

**9.** 1. E or refractory period.    2. B or depolarization.    3. C or polarized.    4. F or repolarization.    5. A or action potential.    6. D or potassium ions.    7. H or sodium-potassium pump.

**10.** 1. A or somatic reflex(es).    2. B or autonomic reflex(es).    3. A or somatic reflex(es).    4. B or autonomic reflex(es).    5. A or somatic reflex(es).    6. B or autonomic reflex(es).    7. B or autonomic reflex(es).

**11.** 1. Pin-prick pain.    2. Skeletal muscle.    3. Two (third with muscle).

**Figure 7–2:**

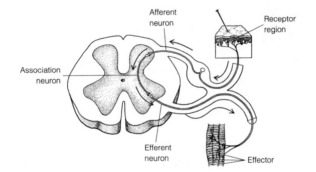

**12.** 1. Neurons.    2. $K^+$ enters the cell.    3. Unmyelinated.    4. Voluntary act.    5. Microglia.    6. Stretch.
7. High $Na^+$.

## Central Nervous System

**13.** 1. Cerebral hemispheres.    2. Brain stem.    3. Cerebellum.    4. Ventricles.    5. Cerebrospinal fluid.

**14.** Circle: Cerebral hemispheres; Cerebellum; Diencephalon.

**15.** 1. Gyrus.    2. Surface area.    3. Neuron cell bodies and unmyelinated fibers.    4. Myelinated fibers.
5. Basal nuclei.

**16. Figure 7–3:** 1. D.    2. L.    3. F.    4. C.    5. K.    6. B.    7. E.    8. A.    9. I.    10. H.    11. J.    12. G.
Areas B and C should be striped.

**17. Figure 7–4:** 1. J.    2. L.    3. O.    4. M.    5. B.    6. D.    7. A.    8. K.    9. G.    10. I.    11. E.    12. N.
13. F.    14. H.    15. C.    Structures #4, #6, #10, and #14 should be blue. Structure #2, the cavity enclosed by #15, #2, and #8, and the entire gray area around the brain should be colored yellow.

**18.** 1. Hypothalamus.    2. Pons.    3. Cerebellum.    4. Thalamus.    5. Medulla oblongata.    6. Corpus callosum.
7. Cerebral aqueduct.    8. Thalamus.    9. Choroid plexuses.    10. Cerebral peduncle.    11. Hypothalamus.

**19.** 1. Postcentral.    2. Temporal.    3. Frontal.    4. Broca's.    5. Left.    6. *T.*    7. Reticular system.    8. *T.*    9. Alert.

**20.** 1. Dura mater.    2. Pia mater.    3. Arachnoid villi.    4. Arachnoid.    5. Dura mater.

**21. Figure 7-5:**

**22.** 1. Choroid plexuses.    2. Ventricles.    3. Cerebral aqueduct.    4. Central canal.    5. Subarachnoid space.
6. Fourth ventricle.    7. Hydrocephalus.

**23.** 1. E or concussion.    2. F or contusion.    3. D or coma.    4. G or intracranial hemorrhage.    5. B or cerebral
edema.    6. C or CVA.    7. A or Alzheimer's disease.    8. H or multiple sclerosis.    9. I or TIA.

**24.** 1. Foramen magnum.    2. Lumbar.    3. Lumbar tap, or puncture.    4. 31.    5. 8.    6. 12.    7. 5.    8. 5.
9. Cauda equina.

**25.** 1. D or association neurons.    2. B or efferent.    3. A or afferent.    4. B or efferent.    5. A or afferent.
6. C or both afferent and efferent.

**26. Figure 7-6:**

**27.** 1. B or loss of sensory function.    2. A or loss of motor function.    3. C or loss of both sensory and motor function.

**Peripheral Nervous System**

**28. Figure 7-7:**

**29.** 1. Nerve (or fascicle).    2. Mixed.    3. Afferent.

**30.    Figure 7-8:**

**31. Figure 7–9:**

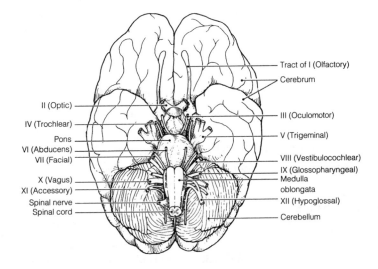

**32.** 1. XI-Accessory.   2. I-Olfactory.   3. III-Oculomotor.   4. X-Vagus.   5. VII-Facial.   6. V-Trigeminal.
7. VIII-Vestibulocochlear.   8. VII-Facial.   9. III, IV, VI.   10. V-Trigeminal.   11. II-Optic.   12. I, II, VIII.

**33.** 1. Plexuses.   2. Limbs and anterolateral body trunk.   3. Thorax.   4. Posterior body trunk.

**34.** 1. Cervical plexus.   2. Phrenic nerve.   3. Sciatic nerve.   4. Fibular and tibial nerves.   5. Median nerve.
6. Musculocutaneous nerve.   7. Lumbar plexus.   8. Femoral nerve.   9. Ulnar nerve.

**35. Figure 7–10:**

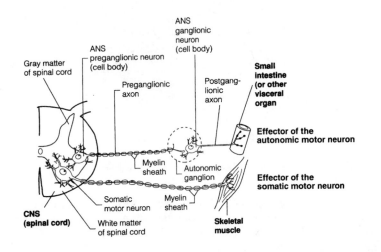

**36.** Check sympathetic for 1, 4, 6, 8, and 10. Check parasympathetic for 2, 3, 5, 7, 9, and 11.

**37.** 1. Increased respiratory rate.   2. Increased heart rate and blood pressure.   3. Increased availability of blood glucose.
4. Pupils dilate; increased blood flow to heart, brain, and skeletal muscles.

## Developmental Aspects of the Nervous System

**38.** 1. Hypothalamus.   2. Oxygen.   3. Cephalocaudal.   4. Gross.   5. Blood pressure.   6. Decreased oxygen
(blood) to brain.   7. Senility.   8. Stroke (CVA).

## The Incredible Journey

**39.** 1. Cerebellum.   2. Medulla.   3. Hypothalamus.   4. Memories.   5. Temporal.   6. Broca's area.   7. Reasoning.
8. Frontal.   9. Vagus (X).   10. Dura mater.   11. Subarachnoid space.   12. Fourth.

## At the Clinic

**40.** Parasympathetic.

**41.** Since the nerve cells are amitotic, the tumor is most likely a glioma, developing from one of the types of neuroglia.

**42.** During sympathetic activation, large amounts of epinephrine from the adrenal medulla pour into the blood. It will
take time for the hormone to be broken down throughout the body.

**43.** Cell bodies of neurons whose axons contribute to the phrenic nerve lie in the C1–C4 segments of the spinal cord. Transection of the cord at this level—at C2 in the first victim of the auto accident—blocks impulses from the respiratory centers in the brain stem that signal these motor neurons, thereby inactivating them and halting the respiratory movements of the diaphragm. This leads to death. Transection of the cord inferior to C1–C4 does not interrupt this respiratory nervous pathway, so the phrenic nerves are still functional and breathing remains normal.

**44.** The stroke has destroyed the trunk, hip, and lower limb region of the primary motor cortex that corresponds to those paralyzed areas on the *left* side of the body. (Remember the motor pathways are crossed.)

**45.** 1. Cerebellum.    2. Basal nuclei.    3. Meningitis.    4. III (oculomotor).    5. Somatosensory cortex.    6. Broca's area. 7. Electroencephalogram.

**46.** Marie has ataxia indicating problems of the cerebellum.

**47.** Paresthesia because the fiber tracts in the dorsal white matter are sensory tracts.

## Chapter 8    Special Senses

### The Eye and Vision

**1.**  1. Extrinsic, or external eye.    2. Eyelids.    3. Meibomian glands.    4. Conjunctivitis.

**2.**  1. 2.    2. 4.    3. 3.    4. 1.

**3.**  **Figure 8–1:** 1. Superior rectus turns eye superiorly.    2. Inferior rectus turns eye inferiorly.    3. Superior oblique turns eye inferiorly and laterally.    4. Lateral rectus turns eye laterally.    5. Medial rectus turns eye medially. 6. Inferior oblique turns eye superiorly and laterally.

**4.**  1. Conjunctiva secretes mucus.    2. Lacrimal glands secrete salt water and lysozyme.    3. Meibomian glands secrete oil. Circle the lacrimal gland secretion.

**5.**  1. L or refraction.    2. A or accommodation.    3. F or emmetropia.    4. H or hyperopia.    5. K or photopupillary reflex.    6. D or cataract.    7. I or myopia.    8. C or astigmatism.    9. G or glaucoma.    10. E or convergence. 11. B or accommodation pupillary reflex.    12. J or night blindness.

**6.**  1. Autonomic nervous system.

**7.**  1. Convex.    2. Real.    3. Behind.    4. Convex (converging).    5. In front of.    6. Concave (diverging).

**8.**  1. L or suspensory ligament.    2. A or aqueous humor.    3. K or sclera.    4. I or optic disk.    5. D or ciliary body. 6. C or choroid coat.    7. B or canal of Schlemm.    8. J or retina.    9. M or vitreous humor.    10. C or choroid coat. 11 and 12. D or ciliary body, G or iris.    13. F or fovea centralis.    14–17. A or aqueous humor, E or cornea, H or lens, and M or vitreous humor.    18. E or cornea.    19. G or iris.

**9.**  **Figure 8–2:**

**10.** 1. In distant vision the ciliary muscle is relaxed, the lens convexity is decreased, and the degree of light refraction is decreased.    2. In close vision the ciliary muscle is contracted, the lens convexity is increased, and the degree of light refraction is increased.

**11.** Retina → Optic nerve → Optic chiasma → Optic tract → Synapse in thalamus → Optic radiation → Optic cortex.

**12.** 1. Three.    2–4. Blue, green, and red.    5. At the same time.    6. Total color blindness.    7. Males.    8. Rods.

**13.** 1. Vitreous humor.    2. Superior rectus.    3. Far vision.    4. Proprioceptors.    5. Iris.

**14.** 1. Opsin.    2. Rhodopsin.    3. Bleaching of the pigment.    4. Yellow.    5. Colorless.    6. A.

## Ear: Hearing and Balance

**15.** 1–3. E, I, and M.    4–6. C, K, and N.    7–9. A, F, and L.    10 and 11. K and N.    12. B.    13. M.    14. C.
15. B.    16 and 17. K and N.    18. G.    19. D.    20. H.

**16. Figure 8–3:** I, E, and M are yellow; A, F, and L are red; C is blue, and K (continuing to N) is green.

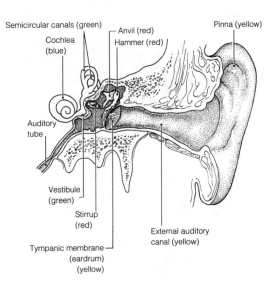

**17.** Eardrum → Hammer → Anvil → Stirrup → Oval window → Perilymph → Membrane → Endolymph → Hair cells.

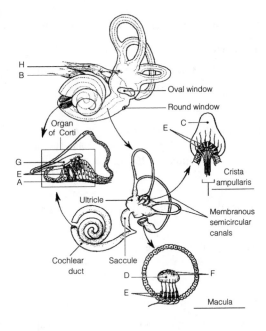

**18. Figure 8–4:**

**19.** 1. C or dynamic.    2. I or semicircular canals.    3. A or angular/rotatory.    4. D or endolymph.    5. B or cupula.
6. J or static.    7 and 8. H or saccule; K or utricle.    9. E or gravity.    10 and 11. G or proprioception; L or vision.

**20.** 1. C.    2. S.    3. S.    4. C.    5. C, S.    6. C.    7. S.

**21.** Nausea, dizziness, and balance problems.

**22.** 1. Pinna.    2. Tectorial membrane.    3. Sound waves.    4. Auditory tube.    5. Optic nerve.

### Chemical Senses: Smell and Taste

**23.** 1–3. (in any order): VII-Facial, IX-Glossopharyngeal; X-Vagus.    4. I-Olfactory.    5. Mucosa of the "roof."
6. Sniffing.    7. Taste buds.    8 or 9. Fungiform; Circumvallate.    10–13. (in any order): Sweet; Salty; Bitter; Sour.
14. Bitter.    15. Smell.    16. Dry.    17. Memories.

**24. Figure 8–5:**

**25. Figure 8–6:**

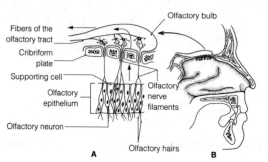

**26.** 1. Musky.    2. Epithelial cell.    3. Neuron.    4. Olfactory nerve.    5. Four receptor types.    6. Metal ions.

### Developmental Aspects of the Special Senses

**27.** 1. Nervous system.    2. Measles (rubella).    3. Blindness.    4. Vision.    5. Hyperopic.    6. Elastic.    7. Presbyopia.
8. Cataract.    9. Presbycusis.

### The Incredible Journey

**28.** 1. Bony labyrinth.    2. Perilymph.    3 and 4. Saccule, Utricle.    5. Gel (otolithic membrane).    6. Otoliths.
7. Macula.    8. Static.    9. Cochlear duct.    10. Organ of Corti.    11. Hearing.    12. Cochlear division of cranial
nerve VIII.    13. Semicircular canals.    14. Cupula.    15. Crista ampullaris.    16. Dynamic.

### At the Clinic

**29.** Patching the strong eye to force the weaker eye muscles to become stronger.

**30.** Cataract; UV radiation, smoking.

**31.** Vision is poor because, without pigment within the eye (in the choroid), light scatters before it can be properly
focused.

**32.** Night blindness; vitamin A; rods.

**33.** The proximal end close to the oval window; sensorineural.

**34.** Otitis externa, most likely because of his exposure to pool bacteria. This diagnosis would be confirmed by presence
of an inflamed external ear canal. If it is otitis media, the middle ear would be inflamed. Bulging of the eardrum
would suggest that inserting ear tubes might be recommended.

**35.** Cranial nerve I, the olfactory nerve.

**36.** Abducens nerve, cranial nerve VI.

**37.** Glaucoma, inadequate drainage of aqueous humor; blindness due to compression of retina and optic disc.

# Chapter 9   The Endocrine System

### The Endocrine System and Hormone Function—An Overview

**1.** 1. F or Slower and more prolonged.   2. E or Nervous system.   3. B or Hormones.   4. D or Nerve impulses.
5. A or Cardiovascular system.

**2.** 1. I or Receptors.   2. N or Target cell(s).   3. A or Altering activity.   4. L or Stimulating new or unusual activities.
5. K or Steroid or amino acid–based.   6. G or Neural.   7. C or Hormonal.   8. D or Humoral.
9. F or Negative feedback.   10. B or Anterior pituitary.   11. J or Releasing hormones.   12. E or Hypothalamus.
13. H or Neuroendocrine.

**3.** Steroid hormones: B, C, D. Amino acid–based hormones: A, E.

### The Major Endocrine Organs

**4. Figure 9–1:**

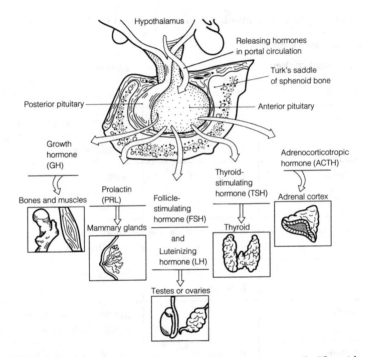

**5. Figure 9–2:** A. Pineal.   B. Posterior pituitary.   C. Anterior pituitary.   D. Thyroid.   E. Thymus.
F. Adrenal cortex.   G. Adrenal medulla.   H. Pancreas.   I. Ovary.   J. Testis.   K. Parathyroids.   L. Placenta.

**6.** 1. C.   2. B.   3–4. F.   5. G.   6. I, L.   7. C.   8. H.   9. H.   10. C.   11. A.   12. B.   13. I, L.
14. C.   15. K.   16. C.   17. J.   18. E.   19. D.   20. D.   21. C.

**7.** 1. Thyroxine.   2. Thymosin.   3. PTH   4. Cortisol (glucocorticoids).   5. Epinephrine.   6. Insulin.
7–10. (in any order): TSH; FSH; LH; ACTH.   11. Glucagon.   12. ADH.   13 and 14. (in any order): FSH; LH.
15 and 16. (in any order): Estrogens; Progesterone.   17. Aldosterone.   18 and 19. (in any order): Prolactin
and oxytocin.

**8.** 1. Estrogen/testosterone.   2. PTH.   3. ADH.   4. Thyroxine.   5. Thyroxine.   6. Insulin.
7. Growth hormone.   8. Estrogen/progesterone.   9. Thyroxine.

**9.** 1. Growth hormone.   2. Thyroxine.   3. PTH.   4. Glucocorticoids.   5. Growth hormone.
6. Androgens (testosterone).

**10.** 1. Polyuria—high sugar content in kidney filtrate causes large amount of water to be lost in the urine.
2. Polydipsia—thirst due to large volumes of urine excreted.   3. Polyphagia—hunger because blood sugar
cannot be used as a body fuel even though blood levels are high.

11. **Figure 9–3:**

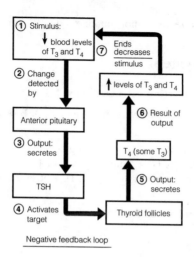

Negative feedback loop

12. 1. Anterior lobe. 2. Steroid hormone. 3. Cortisol. 4. Increases blood $Ca^{2+}$. 5. Growth hormone.

## Other Hormone-Producing Tissues and Organs

13.

| Hormone | Chemical makeup | Source | Effects |
|---|---|---|---|
| Gastrin | Peptide | Stomach | Stimulates stomach glands to secrete HCl |
| Secretin | Peptide | Duodenum | Stimulates the pancreas to secrete $HCO_3^-$-rich juice and the liver to release more bile; inhibits stomach glands |
| Cholecystokinin | Peptide | Duodenum | Stimulates the pancreas to secrete enzyme-rich juice and the gallbladder to contract; relaxes sphincter of Oddi |
| Erythropoietin | Glycoprotein | Kidney in response to hypoxia | Stimulates production of red blood cells by bone marrow |
| Active vitamin $D_3$ | Steroid | Skin; activated by kidneys | Enhances intestinal absorption of calcium |
| Atrial natriuretic peptide (ANP) | Peptide | Heart | Inhibits $Na^+$ reabsorption by kidneys; inhibits aldosterone release by kidney |
| Human chorionic gonadotropin (hCG) | Protein | Placenta | Stimulates corpus luteum to continue producing estrogens and progesterone preventing menses |

## Developmental Aspects of the Endocrine System

14. 1. Neoplasm. 2. Hypersecretion. 3. Iodine. 4. Estrogens. 5. Menopause. 6. Bear children. 7. Insulin.

## The Incredible Journey

15. 1. Insulin. 2. Pancreas. 3. Posterior pituitary, or hypothalamus. 4. ADH. 5. Parathyroid. 6. Calcium. 7. Adrenal medulla. 8. Epinephrine. 9. Thyroxine.

## At the Clinic

16. Pituitary dwarfs who secrete inadequate amounts of GH have fairly normal proportions; cretins (hypothyroid individuals) retain childlike body proportions.

17. Hypothyroidism; iodine deficiency (treated by dietary iodine supplements) or thyroid cell burnout (treated by thyroid hormone supplements).

18. Adrenal cortex.

**19.** For the giant, GH is being secreted in excess by the anterior pituitary, resulting in extraordinary height. For the dwarf, GH is deficient, resulting in very small stature but normal body proportions. For the fat man, $T_3$ and $T_4$ are not being produced adequately, resulting in depressed metabolism and obesity (myxedema). The bearded lady has a tumor of her adrenal cortex (androgen-secreting area) leading to excessive hairiness (hirsutism).

**20.** Stressor → hypothalamus → CRH (releasing hormone) released to the hypophysial portal system blood → to anterior pituitary which releases ACTH → acts on adrenal cortex to trigger releases of glucocorticoids (cortisol, etc.).

**21.** Prolactin.

# Chapter 10  Blood

## Composition and Functions of Blood

**1.** 1. Connective tissue.  2. Formed elements.  3. Plasma.  4. Clotting.  5. Erythrocytes.  6. Hematocrit.  7. Plasma.  8. Leukocytes.  9. Platelets.  10. One.  11. Oxygen.

**2.** 1. F or neutrophil.  2–4. (in any order): C or eosinophil, D or basophil, F or neutrophil.  5. A or red blood cell.  6 and 7. (in any order): E or monocyte, F or neutrophil.  8 and 9. (in any order): E or monocyte, G or lymphocyte.  10. B or megakaryocyte.  11. H or formed elements.  12. C or eosinophil.  13. D or basophil.  14. G or lymphocyte.  15. A or red blood cell.  16. I or plasma.  17. E or monocyte.  18. D or basophil.  19–23. (in any order): C or eosinophil, D or basophil, E or monocyte, F or neutrophil, G or lymphocyte.

**3.** **Figure 10–1:** 100–120 days.

**4.** **Figure 10–2:** A is a neutrophil; B is a monocyte; C is an eosinophil; D is a lymphocyte.

**5.** 1. Diapedesis.  2. *T.*  3. Kidneys.  4. 7.35.  5. 5.5.  6. *T.*  7. *T.*  8. 4.5–5.5.  9. Hematocrit.  10. Less.  11. Monocytes.  12. Lymphocytes.

**6.** 1. Erythrocytes.  2. Monocytes.  3. Lymphocyte.  4. Platelets.  5. Aneurysm.  6. Hemoglobin.  7. Lymphocyte.

**7.** Check 1, 2, 3.

### Hemostasis

**8.** 1. A or break.    2. E or platelets.    3. G or serotonin.    4. I or thromboplastin.    5. F or prothrombin.
6. H or thrombin.    7. D or fibrinogen.    8. C or fibrin.    9. B or erythrocytes.

**9.** 1. 3–6 min.    2. Heparin.    3. *T.*

### Blood Groups and Transfusions

**10.**

| Blood type | Agglutinogens or antigens | Agglutinins or antibodies in plasma | Can donate blood to type | Can receive blood from type |
|---|---|---|---|---|
| 1. Type A | A | anti-B | A, AB | A, O |
| 2. Type B | B | anti-A | B, AB | B, O |
| 3. Type AB | A, B | none | AB | A, B, AB, O |
| 4. Type O | none | anti-A, anti-B | A, B, AB, O | O |

**11.** Type O is the universal donor. AB is the universal recipient.

**12.** A reaction during which plasma antibodies attach to and lyse red blood cells different from your own.

### Developmental Aspects of Blood

**13.** 1. F.    2. Jaundiced.    3. Sickle-cell.    4. Hemophilia.    5. Iron.    6. Pernicious.    7. $B_{12}$.    8. Thrombi.
9. Leukemia.

### The Incredible Journey

**14.** 1. Hematopoiesis.    2. Hemostasis.    3. Hemocytoblasts.    4. Neutrophil.    5. Phagocyte.    6. Erythropoietin.
7. Red blood cells.    8. Hemoglobin.    9. Oxygen.    10. Lymphocytes.    11. Antibodies.    12–15. (in any order):
Basophils; Eosinophils; Monocytes; Platelets.    16. Endothelium.    17. Platelets.    18. Serotonin.    19. Fibrin.
20. Clot.    21. Thromboplastin.    22. Prothrombin.    23. Thrombin.    24. Fibrinogen.    25. Embolus.

### At the Clinic

**15.** Hemolytic disease of the newborn.

**16.** Its RBCs have been destroyed by the mother's antibodies; therefore the baby's blood is carrying insufficient oxygen.

**17.** She must have received mismatched (Rh$^+$) blood previously in a transfusion.

**18.** Give the mother RhoGAM to prevent her from becoming sensitized to the Rh$^+$ antigen.

**19.** Fetal progress will be followed in expectation of hemolytic disease of the newborn; intrauterine transfusions will be given if necessary, as well as complete blood transfusion to the newborn.

**20.** No; A+.

**21.** The stem cells for hematopoiesis in red bone marrow are a rapidly dividing cell population. Hence, they would be targeted (along with other rapidly dividing cells) by chemotherapeutic drugs.

**22.** Virtually all bones contain red marrow and functional hematopoietic tissue in young children, but in adults only the sternum, ilium, and a very few long bone epiphyses contain red marrow.

## Chapter 11    The Cardiovascular System

### Cardiovascular System: The Heart

**1.** 1. Thorax.    2. Diaphragm.    3. Second.    4. Aorta.    5. Right atrium.    6. Atria.    7. Ventricles.
8. Endocardium.    9. Epicardium.    10. Friction.    11. Cardiac muscle.

**2.** 1. Right ventricle.    2. Pulmonary semilunar.    3. Pulmonary arteries.    4. Lungs.    5. Right and left pulmonary veins.    6. Left atrium.    7. Mitral (bicuspid).    8. Left ventricle.    9. Aortic.    10. Aorta.    11. Capillary beds.
12. Superior vena cava.    13. Inferior vena cava.

In Figure 11–1 on p. 327, the white areas represent regions transporting $O_2$-rich blood. The gray vessels transport $O_2$-poor blood.

**Figure 11–1:**

3. **Figure 11–2:** 1. Right atrium.    2. Left atrium.    3. Right ventricle.    4. Left ventricle.    5. Superior vena cava.
6. Inferior vena cava.    7. Aorta.    8. Pulmonary trunk.    9. Left pulmonary artery.    10. Right pulmonary artery.
11. Right pulmonary veins.    12. Left pulmonary veins.    13. Coronary circulation.    14. Apex of heart.
15. Ligamentum arteriosum.

4. **Figure 11–3:**

5. 1. Systole.    2. Diastole.    3. Lub-dup.    4. Atrioventricular.    5. Semilunar.    6. Ventricles.    7. Atria.
8. Atria.    9. Ventricles.    10. Murmurs.

6. **Figure 11–4:** Red arrows should be drawn from the left atrium to the left ventricle and out the aorta. Blue arrows should be drawn from the superior and inferior vena cavae into the right atrium, then into the right ventricle and out the pulmonary trunk. Green arrows should be drawn from #1 to #5 in numerical order.

1. SA node.    2. AV node.    3. AV bundle or bundle of His.    4. Bundle branches.    5. Purkinje fibers.
6. Pulmonary valve.    7. Aortic valve.    8. Mitral (bicuspid) valve.    9. Tricuspid valve.

A and B. (in any order): 6, 7.    C and D. (in any order): 8, 9.    E. 9.    F. 8.    G. 1.    H. 2.

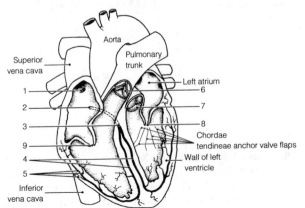

**7.** 1. C or electrocardiogram.    2. F or P wave.    3. H or T wave.    4. G or QRS wave.    5. B or bradycardia.    6. D or fibrillation.    7. I or tachycardia.    8. E or heart block.    9. A or angina pectoris.

**8. Figure 11–5:**

**9.** 1. Cardiac output.    2. Heart rate.    3. Stroke volume.    4. About 75 beats per minute.    5. 70 ml per beat.    6. 5250 ml per minute.    7. Minute.    8. Stroke volume.    9. Stretch.    10. Blood.

**10.** Check 1, 2, 4, 5, 6, 8, and 10.

**11.** 1. Fetal.    2. Rate of contraction.    3. Left.    4. *T.*    5. *T.*

**12.** 1. Left side of heart.    2. P wave.    3. AV valves opened.    4. Aortic semilunar valve.    5. Tricuspid valve.    6. Heart block.

## Cardiovascular System: Blood Vessels

**13.** 1. Lumen.    2. Vasoconstriction.    3. Vasodilation.    4. Veins.    5. Arteries.    6. Arterioles.    7. Venules.

**14.** Arteries are high-pressure vessels. Veins are low-pressure vessels. Blood flows from high to low pressure. The venous valves help to prevent the backflow of blood that might otherwise occur in those low-pressure vessels.

**15.** Skeletal muscle activity and breathing (respiratory pump).

**16.** 1. A or tunica intima.    2. B or tunica media.    3. A or tunica intima.    4. A or tunica intima.    5. C or tunica externa.    6. B or tunica media.    7. C or tunica externa. **Figure 11–6:** A. Artery; thick media; small, round lumen. B. Vein; thin media; elongated, relatively collapsed lumen; a value present. C. Capillary; single layer of endothelium. In A and B, the tunica intima is the innermost vessel layer; the tunica externa is the outermost layer; and the tunica media is the thick middle layer.

**17. Figure 11–7:**                    **Figure 11–8:**

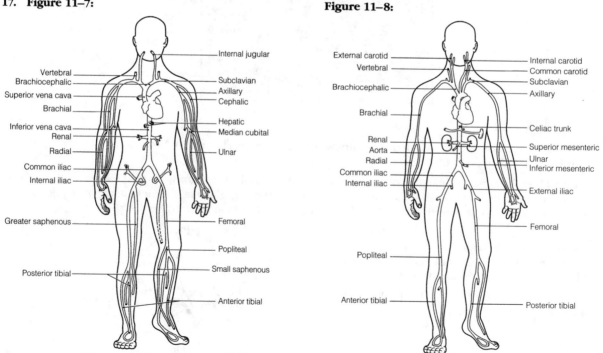

**18.** 1 and 2. S or radial and X or ulnar.    3. U or subclavian.    4. E or cardiac.    5. T or renal.    6. Q or internal jugular.    7. D or brachiocephalic.    8 and 9. A or anterior tibial and R or posterior tibial.    10. M or hepatic portal.    11. F or cephalic.    12. J or gonadal.    13. B or azygos.    14. O or inferior vena cava.    15. L or hepatic.    16–18. I or gastric, N or inferior mesenteric, and V or superior mesenteric.    19. K or great saphenous.    20. G or common iliac.    21. H or femoral.

**19. Figure 11–9:**

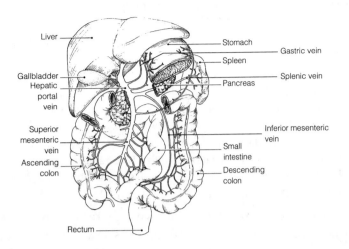

Liver — Stomach — Gastric vein — Spleen — Splenic vein — Gallbladder — Hepatic portal vein — Pancreas — Superior mesenteric vein — Inferior mesenteric vein — Small intestine — Ascending colon — Descending colon — Rectum

**20.** 1 and 2. F or common carotid and W or subclavian.    3. H or coronary.    4 and 5. P or internal carotid and Y or vertebral.    6. B or aorta.    7. J or dorsalis pedis.    8. I or deep femoral.    9. S or phrenic.    10. C or brachial. 11. C or brachial.    12. N or inferior mesenteric.    13. Q or internal iliac.    14. L or femoral.    15. C or brachial. 16. X or superior mesenteric.    17. G or common iliac.    18. E or celiac trunk.    19. K or external carotid. 20–22. (in any order): A or anterior tibial, R or peroneal, T or posterior tibial.    23. U or radial.

**21. Figure 11–10:**

Frontal lobe of cerebral hemisphere — Communicating branch — Anterior cerebral arteries — Middle cerebral artery — Pituitary gland — Internal carotid artery — Communicating branches — Posterior cerebral artery — Pons — Basilar artery — Vertebral artery — Cerebellum

**22. Figure 11–11:**

Superior vena cava — Ductus venosus — Liver — Hepatic portal vein — Umbilical vein — Umbilical cord — Umbilical arteries — Inferior vena cava — Aorta — Common iliac artery — Internal iliac artery — Fetal bladder

Ductus arteriosus — Foramen ovale — Ductus venosus — Inferior vena cava

**Enlarged view of heart**

**23.** 1. C or circle of Willis.   2. J or umbilical vein.   3. E or ductus venosus.   4. A or anterior cerebral artery, G or middle cerebral artery.   5. B or basilar artery.   6. D or ductus arteriosus.   7. F or foramen ovale.

**24.** The fetal lungs are not functioning in gas exchanges, and they are collapsed. The placenta makes the gas exchanges with the fetal blood.

**25.** 1. Vein.   2. Carotid artery.   3. Vasodilation.   4. High blood pressure.   5. Vasodilation.

**26.** 1. H or pulse.   2. B or blood pressure.   3 and 4. C or cardiac output and F or peripheral resistance.   5. D or constriction of arterioles.   6. J or systolic blood pressure.   7. E or diastolic blood pressure.   8. A or over arteries.   9. G or pressure points.   10. I or sounds of Korotkoff.

**27.** 1. G or interstitial fluid.   2. C or diffusion.   3. E or fat soluble.   4–6. (in any order): B or capillary clefts; D or fenestrations; I or vesicles.   7. D or fenestrations.   8–9. B or capillary clefts; D or fenestrations.   10. F or hydrostatic pressure.   11. H or osmotic pressure.   12. G or interstitial fluid.   13. A or blood.

**28.** 1. D.   2–5. I.   6. D.   7. D.   8. I.   9–11. D.   12. I.   13. I.

**29.** 1. Femoral artery.   2. Brachial artery.   3. Popliteal artery.   4. Facial artery.   5. Radial artery.

**30.** 1. Increase.   2. Orthostatic.   3. Brain.   4. Stethoscope.   5. Low.   6. *T.*   7. Vasoconstricting.   8. Hypertension.

### Developmental Aspects of the Cardiovascular System

**31.** 1. Fourth.   2. Lungs.   3. Ductus venosus.   4. Umbilical vein.   5. Placenta.   6. Fetal liver.   7. Umbilical arteries.   8. Occluded.   9. Deaths.   10. Atherosclerosis.   11. Menopause.   12. Aerobic exercise.   13. Atherosclerosis.   14. Varicose veins.   15 and 16. Feet and legs.

### The Incredible Journey

**32.** 1. Left atrium.   2. Left ventricle.   3. Mitral.   4. Chordae tendineae.   5. Diastole.   6. Systole/contraction.   7. Aortic semilunar.   8. Aorta.   9. Superior mesenteric.   10. Endothelial.   11. Superior mesenteric vein.   12. Splenic.   13. Nutrients.   14. Phagocytic (Kupffer).   15. Hepatic.   16. Inferior vena cava.   17. Right atrium.   18. Pulmonary trunk.   19. Pulmonary.   20. Lungs.   21. Subclavian.

### At the Clinic

**33.** Zero; myocardial infarction. The posterior interventricular artery supplies much of the left ventricle, the systemic pump.

**34.** Bradycardia, which results from excessive vagal stimulation of the heart, can be determined by taking the pulse.

**35.** Peripheral congestion due to right heart failure.

**36.** Thrombosis, atherosclerosis; an arterial anastomosis (circle of Willis), e.g. (1) Left internal carotid artery to left anterior cerebral artery. Then through anterior communicating branch to right anterior cerebral artery and (2) vertebral arteries to basilar artery to right posterior cerebral artery through the posterior communicating branch to right middle cerebral artery.

**37.** High; polycythemia increases blood viscosity (thus peripheral resistance), which increases blood pressure.

**38.** The stiffened valve flaps would not close properly and the valve would become incompetent. A heart murmur would be heard after the valve had (supposedly) closed and blood was backflowing through the valve.

**39.** Thrombophlebitis occurs when a thrombus (clot) forms in an inflamed blood vessel (a vein). The danger is that the clot may detach, leading to a pulmonary embolism.

**40.** An ECG only reveals electrical problems. It is not useful for revealing valvular problems.

**41.** If anything, exercise extends life by making the cardiovascular and respiratory systems more efficient. Heart rate drops and stroke volume increases.

**42.** When the environmental temperature is high, blood vessels serving the skin vasodilate and much of the blood supply will be found in dermal blood vessels. Then when you stand suddenly, there will initially be inadequate blood volume in the larger, more central blood vessels to ensure that the brain receives a normal blood supply; thus the dizziness.

# Chapter 12   The Lymphatic System and Body Defenses

**The Lymphatic System**

**1.**   1. Pump.   2. Arteries.   3. Veins.   4. Valves.   5. Lymph.   6. 3 liters.

**2.**   **Figure 12–1:**

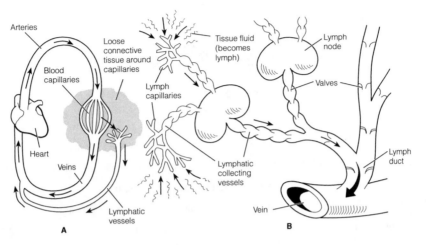

**3.**   1. Blood capillary.   2. Abundant supply of lymphatics.   3. High-pressure gradient.   4. Impermeable.

**4.**   1. C or spleen.   2. A or lymph nodes.   3. D or thymus.   4. B or Peyer's patches, E or tonsils.   5. C or spleen.
6. B or Peyer's patches.

**5.**   **Figure 12–2:** Shade in the right upper limb and right side of the thorax and head.

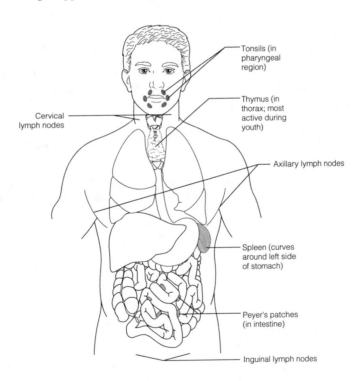

**6.**   1. B lymphocytes.   2. They produce and release antibodies.   3. T lymphocytes.   4. Macrophages; Phagocytes.
5. This slows the flow of lymph through the node, allowing time for immune cells and macrophages to respond to
foreign substances present in the lymph.   6. Valves in the afferent and efferent lymphatics.   7. Cervical, axillary,
inguinal.   8. They act to protect the body by removing bacteria or other debris from the lymphatic stream.

**Figure 12–3:**

**Body Defenses**

**7.** 1. Surface membrane barriers, mucosae.   2. Natural killer cells.   3. Chemicals (inflammatory and antimicrobial).

**8.** 1. Tears and saliva.   2. Stomach and female reproductive tract.   3. Sebaceous (oil) glands; skin.   4. Digestive.

**9.** 1. Itching.   2. Natural killer cells.   3. Interferon.   4. Inflammation.   5. Antibacterial.

**10.** 1. B or lysozyme, F or sebum.   2. C or mucosae, G or skin.   3. A or acids, B or lysozyme, D or mucus, E or protein-digesting enzymes, F or sebum.   4. D or mucus.   5. A–G.

**11.** They propel mucus laden with trapped debris superiorly away from the lungs to the throat, where it can be swallowed or spat out.

**12.** Phagocytosis is ingestion and destruction of particulate material by certain cells. The rougher the particle, the more easily it is ingested.

**13.** Check 1, 3, 4.

**14.** 1. Prevents the spread of damaging agents to nearby areas.   2. Disposes of cell debris and pathogens. 3. Sets the stage for repair (healing).

**15.** 1. F or increased blood flow.   2. E or histamine.   3. G or inflammatory chemicals.   4. A or chemotaxis. 5. C or edema.   6. H or macrophages.   7. B or diapedesis.   8. I or neutrophils.   9. D or fibrin mesh.

**16.** 1. Proteins.   2. Activated.   3. Holes or lesions.   4. Water.   5. Lysis.   6. Opsonization.

**17.** Interferon is synthesized in response to viral infection of a cell. The cell produces and releases interferon proteins, which diffuse to nearby cells, where they prevent viruses from multiplying within those cells.

**18.** 1. Immune system.   2. Proteins.   3. Haptens.   4. Nonself.

**19.** 1. A or antigens.   2. E or humoral immunity.   3. D or cellular immunity.   4 and 5. B or B cells and I or T cells. 6. H or macrophages.   7 and 8. C or blood and F or lymph.   9. G or lymph nodes.

**20. Figure 12–4:**

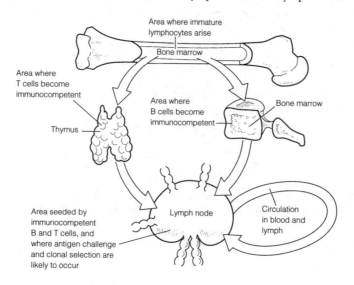

1. The appearance of antigen-specific receptors on the membrane of the lymphocyte.   2. Fetal life.   3. Its genes. 4. Binding to "its" antigen.   5. "Self."

**21.**

| Characteristic | T cell | B cell |
|---|---|---|
| Originates in bone marrow from stem cells called hemocytoblasts | √ | √ |
| Progeny are plasma cells | | √ |
| Progeny include suppressors, helpers, and killers | √ | |
| Progeny include memory cells | √ | √ |
| Is responsible for directly attacking foreign cells or virus-infected cells | √ | |
| Produces antibodies that are released to body fluids | | √ |
| Bears a cell-surface receptor capable of recognizing a specific antigen | √ | √ |
| Forms clones upon stimulation | √ | √ |
| Accounts for most of the lymphocytes in the circulation | √ | |

**22.**  1. Lymphokines.   2. Hapten.   3. Liver.

**23.  Figure 12–5:**

1. The V portion.   2. The C portion.

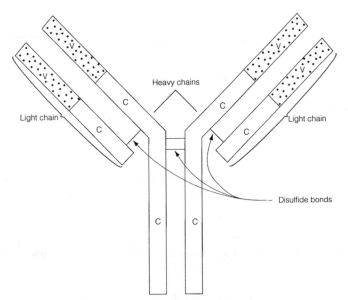

**24.**  1. B or IgD.   2. D or IgG.   3. E or IgM.   4. D or IgG, E or IgM.   5. E or IgM.   6. D or IgG.
7. C or IgE.   8. A or IgA.

**25.**  1. Antigen.   2. Complement activation and lysis.   3. Neutralization.   4. Agglutination.   5. IgM.
6. Precipitation.   7. Phagocytes.

**26.**  1. A.   2–4. P.   5. A.   6. A.

**27.**  1. P.   2. P.   3. S.   4. P.   5. S.

**28.**  1. A or helper T cell.   2. A or helper T cell.   3. C or suppressor T cell.   4. B or killer T cell.

**29.**  1. G or interferon.   2. C or chemotaxis factors.   3. B or antibodies.   4. F or inflammation.
5. E or cytokines.   6. D or complement.   7. E or cytokines.

**30.**  1. Allografts; an unrelated person.   2. Cytotoxic (killer) T cells and macrophages.   3. To prevent rejection, the recipient's immune system must be suppressed. The patient is unprotected from foreign antigens, and bacterial or viral infection is a common cause of death.

**31. Figure 12–6:**

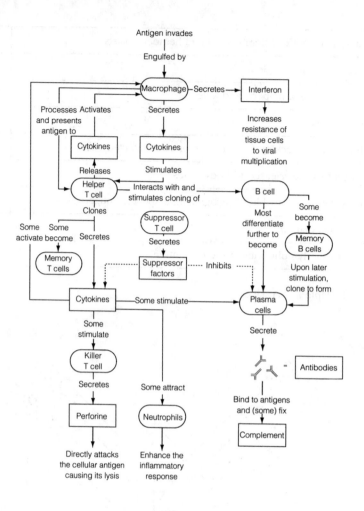

**32.** 1. C or immunodeficiency.    2. A or allergy.    3. A or allergy.    4. C or immunodeficiency.    5. B or autoimmune disease.    6. C or immunodeficiency.    7. B or autoimmune disease.    8. A or allergy.    9. A or allergy.

## Developmental Aspects of the Lymphatic System and Body Defenses

**33.** 1. Veins.    2. Thymus.    3. Spleen.    4. Thymosin.    5. Liver.    6. Lymphatic organs.    7. Birth (or shortly thereafter).    8. Declines.    9–11. (in any order): Immunodeficiencies; Autoimmune diseases; Cancer.    12. IgA.

## The Incredible Journey

**34.** 1. Protein.    2. Lymph node.    3. B lymphocytes (B cells).    4. Plasma cell.    5. Antibodies.    6. Macrophage.    7. Antigens.    8. Antigen presenters.    9. T.    10. Clone.    11. Immunologic memory.

## At the Clinic

**35.** Anaphylactic shock (histamine caused bodywide loss of fluid from the bloodstream); epinephrine injections.

**36.** Contact dermatitis (delayed hypersensitivity) probably caused by a reaction to the chemicals in the detergent used to launder the diapers.

**37.** James is suffering from AIDS. To date, it is nearly 100% fatal.

**38.** She has the classic signs of hypothyroidism (probably due to neck trauma) and she appears to be exhibiting an autoimmune reaction to formerly "hidden antigens" in the thyroid gland colloid.

**39.** Hemorrhage; the spleen is a blood reservoir. No; the liver, bone marrow, and other tissues can take over the spleen's functions.

**40.** The acidity of the vaginal tract inhibits bacterial growth. Hence, anything that decreases vaginal acidity provides an opportunity for bacterial proliferation and vaginal inflammation.

**41.** Lymphedema or swelling due to an accumulation of tissue fluid (lymph) in the area. No, the lymphatic vessels will eventually be replaced by budding from the veins in the area.

# Chapter 13   The Respiratory System

**Functional Anatomy**

**1.** 1. R.    2. L.    3. R.

**2.** 1. External nares or nostrils.    2. Nasal septum.    3–5. (in any order): Warm; Moisten; Cleanse.    6. Paranasal sinuses.    7. Speech.    8. Pharynx.    9. Larynx.    10. Tonsils.    11. Cartilage.    12. Pressure.    13. Anteriorly. 14. Thyroid.    15. Vocal folds or true vocal cords.    16. Speak.

**3.** 1. Mandibular.    2. Alveolus.    3. Larynx.    4. Peritonitis.    5. Nasopharynx.    6. Primary bronchus.

**4.** **Figure 13–1:** In color coding, the pharynx includes the nasopharynx, oropharynx, and laryngopharynx. The larynx runs from the laryngopharynx through the vocal folds to the trachea. The paranasal sinuses include the frontal and sphenoidal sinuses.

**5.** 1. B or bronchioles.    2. G or palate.    3. I or phrenic nerve.    4. E or esophagus.    5. D or epiglottis. 6. K or trachea.    7. A or alveoli.    8. H or parietal pleura.    9. L or visceral pleura.    10. F or glottis. 11. C or conchae.

**6.** 1. Elastic connective.    2. Gas exchange.    3. Surfactant.    4. Reduce the surface tension.

**7.** 1. Provides a patent airway; serves as a switching mechanism to route food into the posterior esophagus; acts in voice production (contains vocal folds).    2. Elastic.    3. Hyaline.    4. The epiglottis has to be flexible to be able to flap over the glottis during swallowing. The more rigid hyaline cartilages support the walls of the larynx. 5. Adam's apple.

**Figure 13–2:**

**8.** **Figure 13–3:**

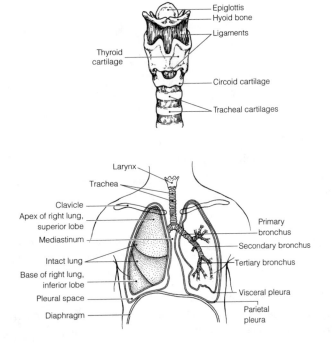

9. **Figure 13–4:** The intact alveoli are the sac-like structures resembling grapes in part A; these should be colored yellow. The small vessels that appear to be spider webbing over their outer surface are the pulmonary capillaries. $O_2$ should be written inside the alveolar chamber and its arrow should move from the alveolus into the capillary. $CO_2$ should be written within the capillary and its arrow shown going from the capillary into the alveolar chamber.

## Respiratory Physiology

10. 1. C or intrapleural pressure.   2. A or atmospheric pressure.   3 and 4. B or intrapulmonary pressure.
    5. C or intrapleural pressure.   6. B or intrapulmonary pressure.

11. When the diaphragm is contracted, the internal volume of the thorax increases, the internal pressure in the thorax decreases, the size of the lungs increases, and the direction of air flow is into the lungs. When the diaphragm is relaxed, the internal volume of the thorax decreases, the internal pressure in the thorax increases, the size of the lungs decreases, and the direction of air flow is out of the lungs.

12. 1. C or inspiration.   2. D or internal respiration.   3. E or ventilation.   4. A or external respiration.

13. 1. Transversus abdominis and external and internal obliques.   2. Internal intercostals and latissimus dorsi.

14. 1. Hiccup.   2. Cough.   3. Sneeze.   4. Yawn.

15. 1. E or tidal volume.   2. A or dead space volume.   3. F or vital capacity.   4. D or residual volume.
    5. B or expiratory reserve volume.

16. **Figure 13–5:**

17. 1. F.   2. G.   3. H.   4. B.   5. E.   6. J.   7. D.   8. C.   9. I.

18. 1. Hemoglobin.   2. Bicarbonate ions.   3. Plasma.   4. Oxygen.

19. 1. Acidosis.   2. ↑pH.   3. Hyperventilation.   4. ↑Oxygen.   5. ↑$CO_2$ in blood.   6. ↑$PCO_2$.

20. 1. E or pons centers.   2. C or inspiratory center.   3. F or stretch receptors in lungs.   4. A or chemoreceptors.
    5. B or intercostal, D or phrenic.

## Respiratory Disorders

21. 1. A or apnea.   2. F or eupnea.   3. D or dyspnea.   4. G or hypoxia.   5. E or emphysema.   6. C or chronic bronchitis.   7. B or asthma.   8. C or chronic bronchitis, E or emphysema.   9. H or lung cancer.
    10. I or tuberculosis.

## Developmental Aspects of the Respiratory System

22. 1. Infant respiratory distress syndrome.   2. Surfactant.   3. Lower the surface tension of the watery film in the alveolar sacs.   4. It keeps the lungs inflated so that gas exchange can continue.

**23.** 1. 40.  2. 12–18.  3. Asthma.  4. Chronic bronchitis.  5. Emphysema or TB.  6. Elasticity.
7. Vital capacity.  8. Respiratory infections, particularly pneumonia.

**The Incredible Journey**

**24.** 1. Nasal conchae.  2. Pharyngeal tonsils.  3. Nasopharynx.  4. Mucus.  5. Vocal fold.  6. Larynx.
7. Digestive.  8. Epiglottis.  9. Trachea.  10. Cilia.  11. Throat (pharynx).  12. Primary bronchi.
13. Left.  14. Bronchiole.  15. Alveolus.  16. Red blood cells.  17. Red.  18. Oxygen.  19. Carbon dioxide.
20. Cough.

**At the Clinic**

**25.** Pleurisy.

**26.** Michael most likely is suffering from carbon monoxide poisoning.

**27.** Sudden infant death syndrome (SIDS).

**28.** Chronic bronchitis; smoking inhibits ciliary action.

**29.** Atelectasis. The lungs are in separate pleural cavities so only the left lung will collapse.

**30.** The mucus secreted by the respiratory mucosa will be abnormally thick and difficult to clear. As a result, respiratory passages will become blocked with mucus, which favors respiratory infections.

**31.** The pharyngeal tonsils, which lie at the dorsal aspect of the nasal cavity.

**32.** 1. Hemoglobin, which transports most of the oxygen carried in the blood, is found in RBCs, so if the number of RBCs drops, the amount of oxygen transported drops as well.  2. The mucus increases the thickness of the respiratory membrane, impairing the efficiency of gas diffusion and exchange.  3. One gas in cigarette smoke is carbon monoxide, which competes with oxygen for binding sites on hemoglobin. Also smoking paralyzes the cilia, increasing the patient's risk of passageway obstruction by mucus and infection.

# Chapter 14  The Digestive System and Body Metabolism

**Anatomy of the Digestive System**

**1.** 1. Oral cavity.  2. Digestion.
3. Blood.  4. Eliminated or excreted.
5. Feces.  6. Alimentary canal or GI tract.  7. Accessory.

**2. Figure 14–1:** The ascending, transverse, descending, and sigmoid colon are all part of the large intestine. The parotid, sublingual, and submandibular glands are salivary glands.

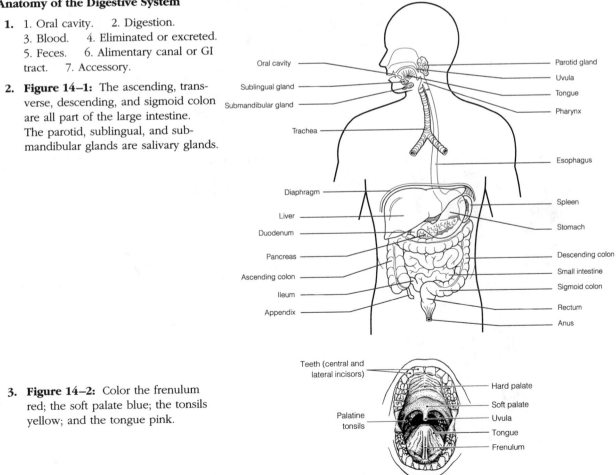

**3. Figure 14–2:** Color the frenulum red; the soft palate blue; the tonsils yellow; and the tongue pink.

**4.** 1. B or intestinal glands.   2. E or salivary glands.   3. D or pancreas.   4. C or liver.   5. A or gastric glands.

**5.** 1. J or mesentery.   2. X or villi.   3. N or Peyer's patches.   4. P or plicae circulares.   5. L or oral cavity, U or stomach.   6. V or tongue.   7. O or pharynx.   8. E or greater omentum, I or lesser omentum, J or mesentery.   9. D or esophagus.   10. R or rugae.   11. G or haustra.   12. K or microvilli.   13. H or ileocecal valve.   14. S or small intestine.   15. C or colon.   16. W or vestibule.   17. B or appendix.   18. U or stomach.   19. I or lesser omentum.   20. S or small intestine.   21. Q or pyloric sphincter.   22. T or soft palate.   23. S or small intestine.   24. M or parietal peritoneum.   25. A or anal canal.   26. F or hard palate.

**6. Figure 14–3:** On part B, the parietal cells should be colored red and the chief cells colored blue.

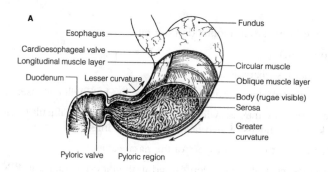

**7.** 1. Esophagus.   2. Rugae.   3. Gallbladder.   4. Cecum.   5. Circular folds.   6. Frenulum.   7. Palatine.   8. Saliva.   9. Protein absorption.

**8. Figure 14–4:** 1. Mucosa.   2. Muscularis externa.   3. Submucosa.   4. Serosa.

**9. Figure 14–5:**

**10. Figure 14–6:**

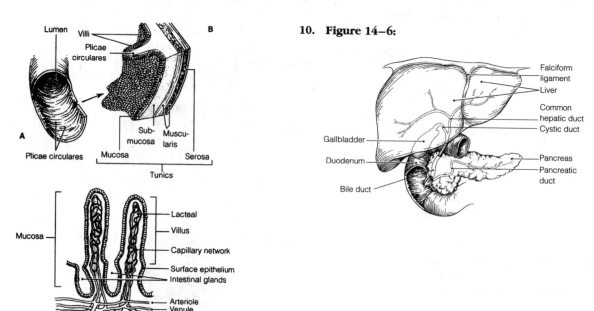

**11.** 1. Deciduous.   2. 6 months.   3. 6 years.   4. Permanent.   5. 32.   6. 20.   7. Incisors.   8. Canine.
9. Premolars.   10. Molars.   11. Wisdom.

**12. Figure 14–7:**

## Functions of the Digestive System

**13.** 1. D or eating.   2. G or swallowing, H or segmentation and peristalsis.   3. E or chewing, F or churning.
4. B or enzymatic breakdown.   5. A or transport of nutrients from lumen to blood.   6. C or elimination of feces.

**14.** 1. O or salivary amylase.   2. G or hormonal stimulus.   3. M or psychological stimulus.   4. I or mechanical
stimulus.   5. L or pepsin.   6. F or HCl.   7. K or mucus.   8. N or rennin.   9. E or churning.   10. C or brush
border enzymes.   11. A or bicarbonate-rich fluid.   12. H or lipases.   13. B or bile.

**15.** 1. G or peritonitis.   2. E or heartburn.   3. F or jaundice.   4. H or ulcer.   5. C or diarrhea.   6. D or gallstones.
7. B or constipation.

**16.** 1. A or cholecystokinin, C or secretin.   2. B or gastrin.   3. A or cholecystokinin.   4. C or secretin.

**17.** 1. C or fructose, D or galactose, E or glucose.   2. F or lactose, G or maltose, I or sucrose.   3. A or amino acids.
4. B or fatty acids.   5. E or glucose.

**18.** 1. P.   2. A.   3. A.   4. P.   5. A.   Circle fatty acids.

**19.** 1. Deglutition.   2. Buccal.   3. Pharyngeal-esophageal.   4. Tongue.   5. Uvula.   6. Larynx.   7. Epiglottis.
8. Peristalsis.   9. Cardioesophageal.   10 and 11. Peristalsis; Segmental.   12. Segmental.   13. Mass movements.
14. Rectum.   15. Defecation.   16. Emetic.   17. Vomiting.

## Nutrition and Metabolism

**20.** 1. B or carbohydrates.   2. C or fats.   3. A or amino acids.   4. C or fats.   5. C or fats.   6. A or amino acids.

**21.** 1. A or bread/pasta, D or fruits, H or vegetables.   2. B or cheese/cream.   3. G or starch.   4. C or cellulose.
5. B or cheese/cream, E or meat/fish.   6. I or vitamins.   7. F or minerals.

**22. Figure 14–8:**

1. Glycolysis does not require oxygen.   2. Krebs cycle and the electron transport chains are oxygen-requiring.
3. In the form of hydrogen atoms bearing high-energy electrons.   4 and 5. The electron transport chain.

**23.** 1. K or Glucose.   2. O or Oxygen.   3. R or Water.   4. H or Carbon dioxide.   5. A or ATP.   6. N or Monosaccharides.   7 and 8. (in any order): C or Acetoacetic acid; D or Acetone.   9. M or Ketosis.   10. I or Essential.
11. F or Ammonia.   12. Q or Urea.

**24.** 1. Albumin.   2. Clotting proteins.   3. Cholesterol.   4. Hyperglycemia.   5. Glycogen.   6. Hypoglycemia.
7. Glycogenolysis.   8. Gluconeogenesis.   9. Detoxification.   10. Phagocytic.   11. Lipoproteins.   12. Insoluble.
13. LDLs.   14. Membranes.   15. Steroid hormones.   16. Liver.   17. Bile salts.   18. Atherosclerosis.
19. A.   20. Iron.

**25.** 1. TMR.   2. ↓ Metabolic rate.   3. Child.   4. Fats.   5. Vasoconstriction.

**26.** 1. D or heat.   2. B or constriction of skin blood vessels, K or shivering.   3. A or blood.   4. F or hypothalamus.
5. J or pyrogens.   6. C or frostbite.   7. H or perspiration, I or radiation.   8. G or hypothermia.
9. E or hyperthermia.

### Developmental Aspects of the Digestive System

**27.** 1. B or alimentary canal.    2. A or accessory organs.    3. D or cleft palate/lip.    4. N or tracheoesophageal fistula.    5. E or cystic fibrosis.    6. H or PKU.    7. K or rooting.    8. M or stomach.    9. C or appendicitis.    10. G or gastritis, O or ulcers.    11. I or periodontal disease.

### The Incredible Journey

**28.** 1. Mucosa.    2. Vestibule.    3. Tongue.    4. Salivary amylase.    5. Peristalsis.    6. Esophagus.    7. Larynx.    8. Epiglottis.    9. Stomach.    10. Mucus.    11. Pepsin.    12. Hydrochloric acid.    13. Pyloric.    14. Lipase.    15. Pancreas.    16. Villi.    17. Ileocecal.

### At the Clinic

**29.** Many vegetables contain incomplete proteins. Unless complete proteins are ingested the value of the dietary protein for anabolism is lost because the amino acids will be oxidized for energy. Beans and grains.

**30.** Heartburn due to a hiatal hernia; esophagitis and esophageal ulcers.

**31.** Heat exhaustion; they should drink a "sports drink" containing electrolytes or lemonade to replace lost fluids.

**32.** Bert has heat stroke. Heavy work in an environment that restricts heat loss results in a spiraling upward of body temperature and cessation of thermoregulation. Bert should be immersed in cool water immediately to bring his temperature down and avert brain damage.

**33.** Diverticula are small herniations of the mucosa through the colon walls, a condition called diverticulosis. They are believed to form when the diet lacks bulk and the volume of residue in the colon is small. The colon narrows and contractions of its circular muscles become more powerful, increasing the pressure on its walls. Diverticulitis is a painful condition in which the diverticula become inflamed. This woman has diverticulitis due to the inflammation of her diverticula.

**34.** Lack of lactase (lactose intolerance); add lactase drops to milk before drinking it.

**35.** Examination of the blood plasma would quickly reveal the presence of lipid breakdown products at above-fasting levels.

**36.** Yo-yo dieting causes dramatic drops in metabolic rate and causes the enzyme that unloads fats from the blood (to be stored in fat deposits) to become much more efficient. Furthermore, if the individual doesn't exercise, when he or she is not dieting, excess calories are stored as fat rather than as muscle, or being used to sustain the higher metabolic rate of muscle tissue.

**37.** Iron. She has hemorrhagic anemia compounded by iron loss.

**38.** Appendicitis is caused by bacteria infection. If untreated, bacterial proliferation may cause the appendix to rupture, resulting in contamination of the peritoneal cavity with feces and life-threatening peritonitis.

## Chapter 15    The Urinary System

**1.** 1. Nitrogenous.    2. Water.    3. Acid-base.    4. Kidneys.    5. Ureters.    6. Peristalsis.    7. Urinary bladder.    8. Urethra.    9. 8.    10. 1½.

### Kidneys

**2. Figure 15–1:**

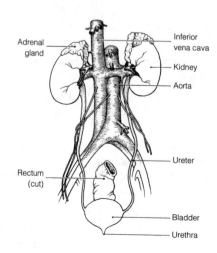

**3. Figure 15–2:** The fibrous membrane surrounding the kidney is the *renal capsule*, the basin-like *pelvis* is continuous with the ureter; a *calyx* is an extension of the pelvis; *renal columns* are extensions of cortical tissue into the medulla. The *cortex* contains the bulk of the nephron structures; the striped-appearing *medullary pyramids* are primarily formed by collecting ducts.

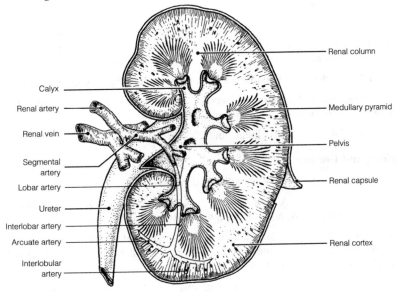

**4.** 1. Intraperitoneal.    2. Urethra.    3. Glomerulus.    4. Glomerulus.    5. Collecting duct.    6. Cortical nephrons.
7. Collecting duct.    8. Glomeruli.

**5. Figure 15–3:** 1. Glomerular capsule.    2. Afferent arteriole.    3. Efferent arteriole.    4. Interlobular artery.
5. Interlobular vein.    6. Arcuate artery.    7. Arcuate vein.    8. Interlobar artery.    9. Interlobar vein.
10. Loop of Henle.    11. Collecting duct.    12. Distal convoluted tubule.    13. Proximal convoluted tubule.
14. Peritubular capillaries.    15. Glomerulus. Relative to the coloring instructions, #1 is green, #15 is red, #14 is
blue, #11 is yellow, and #13 is orange.

**6. Figure 15–4:** Black arrows: Site of filtrate formation is the glomerulus. Arrows leave the glomerulus and enter glomerular (Bowman's) capsule. Red arrows: Major site of amino acid and glucose reabsorption. Shown going *from* the PCT interior and passing through the PCT walls to the capillary bed surrounding the PCT (the latter not shown). Nutrients *leave* the filtrate. Green arrows: At side of ADH action. Arrows (indicating water movement) shown *leaving* the interior of the collecting duct and passing through the walls to enter the capillary bed surrounding that duct. Water *leaves* the filtrate. Yellow arrows: Site of aldosterone action. Arrows (indicating $Na^+$ movement) *leaving* the collecting duct and the DCT, and passing through their walls into the surrounding capillary bed. $Na^+$ *leaves* the filtrate. Blue arrows: Site of tubular secretion. Arrows shown *entering* the PCT to enter the filtrate.

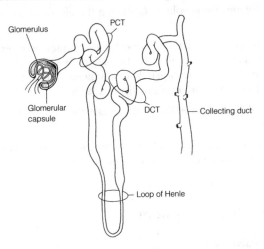

**7.** 1. Afferent.    2. Efferent.    3. Blood plasma.    4 and 5. Diffusion; Active transport.    6. Microvilli.    7. Secretion.
8–10. Diet; Cellular metabolism; Urine output.    11. 1–1.8.    12. Urochrome.    13–15. Urea; Uric acid;  Creatinine.
16. Lungs.    17. Evaporation of perspiration.    18. Decreases.    19. Dialysis.

**8.** 1. Aldosterone. 2. Secretion. 3. ↑K$^+$ reabsorption. 4. ↓BP. 5. ↓K$^+$ retention. 6. ↑HCO$_3^-$ in urine.

**9.** 1. A. 2. B. 3. A. 4. A. 5. B.

**10.** 1. D. 2. D. 3. I. 4. D. 5. I. 6. I.

**11.** 1. L. 2. G. 3. G. 4. A. 5. L. 6. A. 7. G. 8. G. 9. G. 10. A. 11. G. 12. L.

**12.** 1. Hematuria; bleeding in urinary tract. 2. Ketonuria; diabetes mellitus, starvation. 3. Albuminuria; glomerulonephritis, pregnancy. 4. Pyuria; urinary tract infection. 5. Bilirubinuria; liver disease. 6. (No official terminology); kidney stones. 7. Glycosuria; diabetes mellitus.

**13.** 1. All reabsorbed by tubule cells. 2. Usually does not pass through the glomerular filter.

**14.** 1. Chemical buffering; response in less than one second. 2. Adjustment in respiratory rate and depth; response in minutes. 3. Regulation by kidneys; response in hours to days.

**15.** 1. Male adult. 2. Lean adult. 3. Intracellular fluid. 4. Nonelectrolyte. 5. ↑ADH. 6. ↑K$^+$ reabsorption.

### Ureters, Urinary Bladder, and Urethra

**16.** 1. Kidney. 2. Forms urine. 3. Continuous with renal pelvis.

**17.** 1. B or urethra. 2. A or bladder. 3. A or bladder. 4. B or urethra. 5. B or urethra; C or ureter. 6. B or urethra. 7. C or ureter. 8. A or bladder; C or ureter. 9. B or urethra.

**18.** 1. Micturition. 2. Stretch receptors. 3. Contract. 4. Internal urethral sphincter. 5. External urethral sphincter. 6. Voluntarily. 7. About 600. 8. Incontinence. 9. Infants, toddlers. 10 and 11. Emotional/neural problems; Pressure (pregnancy). 12. Urinary retention. 13. Prostate.

**19.** 1. A or cystitis. 2. C or hydronephrosis. 3. F or uremia. 4. E or pyelonephritis. 5. B or diabetes insipidus. 6. D or ptosis.

### Developmental Aspects of the Urinary System

**20.** 1. Placenta. 2. Polycystic disease. 3. Hypospadias. 4. Males. 5. Bladder. 6. 18–24. 7. Glomerulonephritis. 8. Antigen-antibody. 9 and 10. Proteins; Blood. 11. Arteriosclerosis. 12. Tubule. 13 and 14. Urgency; Frequency.

### The Incredible Journey

**21.** 1. Tubule. 2. Renal. 3. Afferent. 4. Glomerulus. 5. Glomerular capsule. 6. Plasma. 7. Proteins. 8. Loop of Henle. 9. Microvilli. 10. Reabsorption. 11 and 12. Glucose; Amino acids. 13. 7.4 (7.35–7.45). 14. Nitrogenous. 15. Sodium. 16. Potassium. 17. Urochrome. 18. Antidiuretic hormone. 19. Collecting duct. 20. Pelvis. 21. Peristalsis. 22. Urine. 23. Micturition. 24. Urethra.

### At the Clinic

**22.** Anuria; renal dialysis.

**23.** Perhaps Eddie is a very heavy sleeper and is thus unresponsive to the "urge" to urinate.

**24.** High sodium content and copious urine volume (although the glucocorticoids can partially take over the role of aldosterone).

**25.** People who are under prolonged stress activate hypothalamic centers that regulate stress by controlling the release of ACTH by the anterior pituitary. Release of ACTH by the anterior pituitary in turn causes both catecholamines and corticosteroids to be released by the adrenal glands to counteract the stressor by raising blood pressure and blood sugar levels. The elevated blood pressure explains his headache.

**26.** The alcohol interferes with the action of ADH, which causes the kidneys to retain water. Hence, excessive body water is being lost in urine.

**27.** Mrs. Rodriques is in a diabetic coma due to lack of insulin. Her blood is acidic and her respiratory system is attempting to compensate by blowing off carbon dioxide (hence, the elevated breathing rate). Her kidneys are reabsorbing bicarbonate.

## Chapter 16   The Reproductive System

### Anatomy of the Male Reproductive System

**1.** Seminiferous tubule → Rete testis → Epididymis → Ductus deferens.

**2.** When body temperature (or external temperature) is high the scrotal muscles relax, allowing the testes to hang lower and farther away from the warmth of the body wall. This causes testicular temperature to drop. When the external temperature is cold, the scrotal muscles contract to draw the testes closer to the body wall.

**3.** 1. E or penis.   2. K or testes.   3. C or ductus deferens.   4. L or urethra.   5. A or bulbourethral glands, G or prostate gland, H or seminal vesicles, K or testes.   6. I or scrotum.   7. B or epididymis.   8. F or prepuce. 9. G or prostate gland.   10. H or seminal vesicles.   11. A or bulbourethral glands.   12. J or spermatic cord.

**4.** **Figure 16–1:** The spongy tissue is the penis; the duct that also serves the urinary system is the urethra; the structure providing ideal temperature conditions is the scrotum; the prepuce is removed at circumcision; the glands producing a secretion that contains sugar are the seminal vesicles; the ductus deferens is cut or cauterized during vasectomy.

**5.** **Figure 16–2:** The site of spermatogenesis is the seminiferous tubule. Sperm mature in the epididymis. The fibrous coat is the tunica albuginea.

## Male Reproductive Functions

**6.** 1. D or spermatogonium.   2. C or secondary spermatocyte, E or sperm, F or spermatid.   3. C or secondary spermatocyte.   4. F or spermatid.   5. E or sperm.   6. A or FSH, G or testosterone.

**Figure 16–3:**

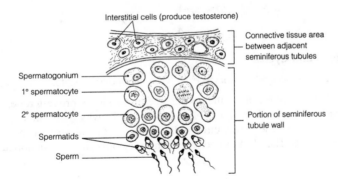

**7.** **Figure 16–4:**

**8.** 1. A or mitosis.    2. B or meiosis.    3. C or both mitosis and meiosis.    4. A or mitosis.    5. B or meiosis. 6. A or mitosis.    7. A or mitosis.    8. B or meiosis.    9. C or both mitosis and meiosis.    10. B or meiosis. 11. B or meiosis.

**9.** Deepening voice; formation of a beard and increased hair growth all over body, particularly in axillary/genital regions; enlargement of skeletal muscles; increased density of skeleton.

### Anatomy of the Female Reproductive System

**10.** 1. Uterus.    2. Vagina.    3. Uterine, or fallopian, tube.    4. Clitoris.    5. Uterine tube.    6. Hymen. 7. Ovary.    8. Fimbriae.

**11.** **Figure 16–5:** The egg travels along the uterine tube after it is released from the ovary. The round ligament helps to anchor the uterus. The ovary produces hormones and gametes.

**12.** **Figure 16–6:** The clitoris should be colored blue; the hymen yellow; and the vaginal opening red.

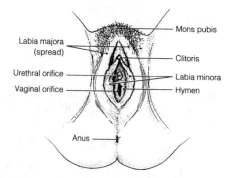

### Female Reproductive Functions and Cycles

**13.** 1. B or primary oocyte.    2. C or secondary oocyte.    3. C or secondary oocyte.    4. D or ovum.

**14.** The follicle (granulosa) cells produce estrogen; the corpus luteum produces progesterone; and oocytes are the central cells in all follicles. Event A = ovulation.    1. No.    2. Peritoneal cavity.    3. After sperm penetration occurs. 4. Ruptured (ovulated) follicle.    5. One ovum; three polar bodies.    6. Males produce four spermatids → four sperm.    7. They deteriorate.    8. They lack nutrient-containing cytoplasm.    9. Menopause.

**Figure 16–7:**

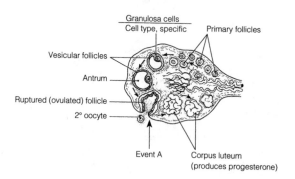

15. Because of this structural condition, many "eggs" (oocytes) are lost in the peritoneal cavity; therefore, they are unavailable for fertilization. The discontinuity also provides infectious microorganisms with access to the peritoneal cavity, possibly leading to PID.

16. 1. FSH (follicle-stimulating hormone).    2. LH (luteinizing hormone).    3. Estrogen and progesterone.
4. Estrogen.    5. LH.    6. LH.

17. Appearance of axillary/pubic hair, development of breasts, widening of pelvis, onset of menses.

18. 1. A or estrogens, B or progesterone.    2. B or progesterone.    3. A or estrogens.    4. B or progesterone.
5 and 6. A or estrogens.

19. **Figure 16–8:** From left to right on part C the structures are: the primary follicle, the growing follicle, the vesicular follicle, the ovulating follicle, and the corpus luteum. In part D, menses is from day 0 to day 4, the proliferative stage is from day 4 to day 14, and the secretory stage is from day 14 to day 28.

## Mammary Glands

20. **Figure 16–9:** The alveolar glands should be colored blue, and the rest of the internal breast, excluding the duct system, should be colored yellow.

## Survey of Pregnancy and Embryonic Development

21. 1. Just its head (the nucleus).    2. Digests away the cement holding the follicle cells together; allows sperm to reach the oocyte.

22. **Figure 16–10:** 1. Fertilization (sperm penetration).    2. Fertilized egg (zygote).    3. Cleavage.    4. Blastocyst (chorionic vesicle).    5. Implantation.

23. 1. H or zygote.    2. F or placenta.    3. B or chorionic villi, C or endometrium.    4. A or amnion.
5. G or umbilical cord.    6. B or chorionic villi.    7. E or fetus.    8. F or placenta.    9. D or fertilization.

24. The blastocyst and then the placenta release hCG, which is like LH and sustains the function of the corpus luteum temporarily.

25. 1. B or mesoderm.    2. C or endoderm.    3. A or ectoderm.    4. B or mesoderm.    5. A or ectoderm.    6. B or mesoderm.    7. C or endoderm.    8. C or endoderm.

26. Oxytocin and prostaglandins.

27. 1. Prolactin.    2. Oxytocin.

28. Check 1, 3, 5, 9, 10, 11, 12.

29. False labor (irregular, ineffective uterine contractions). These occur because rising estrogen levels make the uterus more responsive to oxytocin and antagonize progesterone's quieting influence on the myometrium.

30. 1. Dilation stage: The period from the beginning of labor until full dilation (approx. 10 cm diameter) of the cervix; the longest phase.    2. Expulsion stage: The period from full dilation to the birth (delivery).    3. Placental stage: Delivery of the placenta, which follows delivery of the infant.

**31. Figure 16–11:**

**32.** Each pass forces the baby farther into the birth passage. The cycle ends with the birth of the baby.

**33.** The response to the stimulus enhances the stimulus. For example, the more a baby descends into the pelvis and stretches the uterus, the more oxytocin is produced and the stronger the contractions become.

### Developmental Aspects of the Reproductive System

**34.** 1. Y and X.   2. 2 Xs.   3. Male external genitalia and accessory structures.   4. Female external genitalia and duct system.   5. Cryptorchidism.   6–8. (in any order) *Escherichia coli,* STDs or venereal disease, Yeast infections.   9. PID (pelvic inflammatory disease).   10. Venereal disease microorganisms.   11. Breast.   12. Cervix of the uterus.   13. Pap smear.   14. Menopause.   15. Hot flashes.   16. Declines.   17. Rise.   18. Estrogen.   19. Vaginal.   20. Prostate gland.   21 and 22. Urinary; Reproductive.

### The Incredible Journey

**35.** 1. Uterus.   2. Ovary.   3. Fimbriae.   4. Ovulation.   5. 2° oocyte.   6. Follicle.   7. Peristalsis.   8. Cilia.   9. Sperm.   10. Acrosomes.   11. Meiotic.   12. Ovum.   13. Polar body.   14. Dead.   15. Fertilization.   16. Zygote (fertilized egg).   17. Cleavage.   18. Endometrium.   19. Implantation.   20. Vagina.

### At the Clinic

**36.** Pitocin will act on the placenta, stimulating production and release of prostaglandins. Pitocin and prostaglandins are powerful uterine muscle stimulants. Oxytocin normally causes frequent and vigorous contractions of the uterine wall.

**37.** Megadoses of testosterone would inhibit anterior pituitary gonadotropin (FSH) release. Spermatogenesis is inhibited in the absence of FSH stimulation.

**38.** Her tubes were probably scarred by PID. Hormonal testing and the daily basal temperature recordings would have indicated her anovulatory condition.

**39.** His scrotal muscles had contracted to draw the testes closer to the warmth of the abdominal cavity.

**40.** Mary's fetus might have respiratory problems or even congenital defects due to her smoking, since smoking causes vasoconstriction, which would hinder blood delivery to the placenta.

**41.** Cervical cancer.

# A.D.A.M.® Correlations

This handy reference tool enhances your knowledge of anatomy and physiology when you have access to the *A.D.A.M. The Inside Story*® software program. Organized by chapter, the table below lists where links occur between *A.D.A.M. The Inside Story*® and selected figures from the *Anatomy & Physiology Coloring Workbook* and its companion text, *Essentials of Human Anatomy & Physiology*. Figures are identified by figure number and key structure and are listed with the appropriate view and layer in *A.D.A.M.*®.

Although your instructor will, no doubt, help you to get started with *A.D.A.M. The Inside Story*® here are a couple of hints that will be important in helping you navigate through the program. When in the "Full Anatomy" option, the term "View" refers to the "Anterior," "Lateral," "Posterior," or "Medial" view in the primary window of the computer screen. "Layer" refers to one or more of the layers listed at the bottom of the screen. Using the indicated layer will bring the structure into full view.

| *Coloring Workbook* Figure Number | *Essentials* Figure Number | Key Structures | A.D.A.M.® View and Layer |
|---|---|---|---|
| 1.1 | 1.2f | Cardiovascular system | Using "View" on the Menu Bar, choose Cardiovascular system |
| 1.2 | 1.2h | Respiratory system | Using "View" on the Menu Bar, choose Respiratory system |
| 1.3 | 1.2d | Nervous system | Using "View" on the Menu Bar, choose Nervous system |
| 1.4 | 1.2j | Urinary system | Using "View" on the Menu Bar, choose Urinary system; using "Options," choose Female; using "Options," choose Male |
| 1.5 | 1.2i | Digestive system | Using "View" on the Menu Bar, choose Digestive system |
| 1.6 | 1.2k, l | Reproductive system | Using "View" on the Menu Bar, choose Reproductive system; using "Options," choose Female; using "Options," choose Male |
| 4.1 | 4.1c | Pericardium | Go to "Full Anatomy" View: Anterior Layer: Heart-cut section |

| Coloring Workbook Figure Number | Essentials Figure Number | Key Structures | A.D.A.M.® View and Layer |
|---|---|---|---|
| 4.1 | 4.1c | Peritoneums, serous membranes | Go to "Full Anatomy" View: Anterior Layer: Peritoneum |
| 5.3a | 5.7 | Skull bones | Go to "Full Anatomy" View: Lateral Layer: Skull |
| 5.3c | 5.11 | Skull bones | Go to "Full Anatomy" View: Anterior Layer: Skull |
| 5.4 | 5.10a | Paranasal sinuses | Go to "Full Anatomy" View: Anterior Layer: Skull-coronal section |
| 5.4 | 5.10b | Paranasal sinuses | Go to "Full Anatomy" View: Medial Layer: Nasal cavity |
| 5.6 | 5.14 | Vertebral column | Go to "Full Anatomy" View: Lateral Layer: Sagittal section of bones |
| 5.7 | 5.19 | Thorax | Using "View" on the Menu Bar, choose Skeletal system |
| 5.8 | 5.20 | Scapula | Go to "Full Anatomy" View: Posterior Layer: Scapula |
| 5.13 | 5.1 | Skeletal system | Using "View" on the Menu Bar, choose Skeletal system |
| 5.13 | 5.6 | Skeletal system | Using "View" on the Menu Bar, choose Skeletal System |
| 6.7 and 6.11 | 6.15 and 6.20 | Superficial muscles of the anterior body | Using "View" on the Menu Bar, choose Muscular system |
| 6.8 and 6.12 | 6.16 and 6.21 | Superficial muscles of the posterior body | Using "View" on the Menu Bar, choose Muscular system |
| 6.10 | | Superficial muscles of the anterior forearm | Using "View" on the Menu Bar, choose Muscular system |
| 6.10 | | Superficial muscles of the posterior arm | Go to "Full Anatomy" View: Posterior Layer: Deltoid muscle |

| *Coloring Workbook* Figure Number | *Essentials* Figure Number | Key Structures | A.D.A.M.® View and Layer |
|---|---|---|---|
| 7.3 | 7.13 | Surface of the brain | Go to "Full Anatomy" View: Lateral Layer: Brain |
| 7.4 | 7.15 | Midsagittal brain | Go to "Full Anatomy" View: Medial Layer: Brain and spinal cord |
| 7.5 | 7.16 | Meninges of the brain | Go to "Full Anatomy" View: Anterior Layer: Bones-coronal section |
| 8.1 | 8.2 | Extrinsic muscles of the eye | Go to "Full Anatomy" View: Lateral Layer: Eyeball |
| 9.2 | 9.1 | Endocrine system (Female) | Using "Options," choose Female; using "View" on the Menu Bar, choose Endocrine system |
| 9.2 | 9.1 | Endocrine system (Male) | Using "Options," choose Male; using "View" on the Menu Bar, choose Endocrine system |
| 11.2 | 11.2a | Heart coronary vessels and great vessels | Go to "Full Anatomy" View: Anterior Layer: Cardiac veins |
| 11.4 | 11.2b | Interior of the heart | Go to "Full Anatomy" View: Anterior Layer: Heart-cut section |
| 11.7 | 11.12 | Veins of the systemic circulation | Using "View" on the Menu Bar, choose Cardiovascular system |
| 11.8 | 11.11 | Arteries of the systemic circulation | Using "View" on the Menu Bar, choose Cardiovascular system |
| 11.9 | 11.14 | Hepatic portal circulation | Go to "Full Anatomy" View: Anterior Layer: Portal vein and tributaries |
| 13.1 | 13.2 | Upper respiratory tract | Go to "Full Anatomy" View: Medial Layer: Nasal cavity |
| 13.3 | | Lower respiratory tract | Go to "Full Anatomy" View: Anterior Layer: Tracheal lymph nodes |

| *Coloring Workbook Figure Number* | *Essentials Figure Number* | Key Structures | A.D.A.M.® View and Layer |
|---|---|---|---|
| 14.1 | 14.1 | Digestive system | Using "View" on the Menu Bar, choose Digestive system |
| 14.2 | 14.2a | Oral cavity | Go to "Full Anatomy" View: Anterior Layer: Oral mucosa |
| 14.3 | 14.5 | Stomach | Go to "Full Anatomy" View: Anterior Layer: Stomach-coronal section |
| 14.6 | 14.4 14.6 | Liver, gallbladder, and associated ducts | Go to "Full Anatomy" View: Anterior Layers: Liver, Gallbladder |
| 14.6 | 14.6 | Pancreas and duodenum | Go to "Full Anatomy" View: Anterior Layer: Pancreas |
| 15.1 | 15.1 | Female urinary system; Male urinary system | Using "View" on the Menu Bar, choose Urinary system; using "Options," choose Female; using "Options," choose Male |
| 15.2 | 15.2 | Internal kidney | Go to "Full Anatomy" View: Anterior Layer: Ureter |
| 16.1 | 16.2 | Male reproductive system | Using "Options," choose Male Go to "Full Anatomy" View: Medial Layer: Urinary bladder and prostate gland |
| 16.5 | 16.8 | Female reproductive system | Using "Options," choose Female Go to "Full Anatomy" View: Medial Layer: Uterus |
| 16.9 | 16.13 | Breast | Using "Options," choose Female Go to "Full Anatomy" View: Anterior Layer: Mammary gland |